Luftwaffe Confidential

German aeronautical influence reached the Argentinian Pampas through the IAe – 38, designed by Reimar Horten.
Photo: courtesy Peter F. Selinger.

Luftwaffe Confidential

fundamentals of modern aeronautical design

Perhaps the most fascinating German project of WWII, the Horten IX,
seen here in its glider version during a test flight
Photo: courtesy Jan Scott

Luftwaffe Confidential
Fundamentals of Modern Aeronautical Design - 1st edition - 2012

Text:	Claudio Lamas de Farias
Design and illustrations:	Daniel Uhr
English Translation:	Claudio Lamas de Farias
Revision:	Raul Blacksten
Final editing:	Martin Simons
Cover illustration:	Daniel Uhr
Publisher:	Eqip Werbung & Verlag GmbH
	Sprottauer Str. 52
	53117 Bonn Germany
	phone +49.228.96699011
	fax +49.228.96699012
	www.eqip.de
	eqip@eqip.de
Printed by:	L.E.G.O. SpA, Vicenza, Italy

The authors made all efforts to contact copyright owners of all images in this book. Regrettably not all could be detected properly enough. Therefore the authors apologize for all errors and omissions and kindly ask for corrections for further editions.

© Lamas de Farias, Claudio; Uhr, Daniel; 2012
All rights preserved. No part of this book may be reproduced or transmitted in any form or by any means, electronic or mechanical without the permission of the publisher in righting.

ISBN 978-3-9808838-4-9

Acknowledgments

This book would not have been possible without the help and support of many friends (some old and others recent), companies and institutions who made available many of the photographs. The authors would like to express their most sincere gratitude to all:

Al Bowers
Antonio Almeida Vilaça
Átila de Melo Coutinho (in memoriam)
Bernardo Malfitano - http://www.airshowfan.com
Beth Leal
Comando Geral de Tecnologia Aeroespacial (CTA)
Dan Johnson - http://www.luft46.com
David Ilott
Dan Shumaker
Frank Anthony Barral Dodd
Franz Selinger
Fátima e Isaac Uhr
Hans-Peter Dabrowski
Jan Scott
Klaus Fey
Leila Longo
Leonid Faerberg - http://www.transport-photo.com
Luís Cláudio Portugal Nascimento.
Dr. Manfred Reinhardt
Martin Simons
Mike Freer
Museu Aeroespacial (Rio de Janeiro)
Museo Nacional de Aeronáutica (Morón – Buenos Aires)
NASA
National Museum of the United States Air Force (USAF Museum)
Otávio Lamas de Farias
Parque de Material Aeronáutico do Galeão
Peter F. Selinger
Phil Juvet - http://www.philsaeronauticalstuff.com
Rafael Garcia
Rainer Niedrée
Raul Blacksten
Santiago Rivas
Stevie Beats - http://www.steviebeats.co.uk/
Steve Williams
United States Navy

FOREWORD

Although I was heavily involved in probing and assessing the advanced aviation technology of the Third Reich in the immediate aftermath of World War II, I have never read a more comprehensive listing and commentary of what was actually achieved or planned by Germany's brilliant aeronautical scientists and engineers.

The scope of the innovatory ingenuity that is portrayed is breathtaking, and indeed somewhat sobering if one contemplates what might have been if hostilities had not ended when they did, but had been further prolonged.

What shines through is the Teutonic fascination with rocketry, which after the war was transferred to America along with its top experts in that field, and introduced us to space exploration and captivated the world of science.

Another less publicised activity in which the Germans held a commanding lead was that of VTOL, and most will be surprised to find out how deep they had investigated convertiplanes and coleopter aircraft.

Frankly I found Luftwaffe Confidential a most fascinating read, and a lot of research must have gone into its production. It will be hailed by the large number of aviation historians who find the experimental background to the German Luftwaffe an irresistable attraction, even if somewhat of an enigma.

CAPTAIN ERIC BROWN
C.O. Enemy Aircraft Flight
RAE, Farnborough (1945-7)

CONTENTS

Foreword	09
Introduction	12
1. Aerodynamics	21
Positively swept wings	23
Forward swept wings	28
Crescent wings	31
Variable sweep wings	32
Oblique wings	34
"W" wings	36
Variable incidence wings	38
Delta wings	41
Other aerodynamic innovation	45
Datafiles	47
2. Configuration	57
Tailless aircraft	58
Flying wings	63
Twin fuselage aircraft	70
Asymetric aircraft	72
Butterfly tail	73
"T" tail	78
Further projects in Argentina	82
Ta 183 descendants	85
Canards	90
Datafiles	95
3. VTOL	109
Convertiplanes	113
Coleopter aircraft	120
Datafiles	127
4. Special missions	129
The Packplane	131
Parasite aircraft	133
Spy aircraft	134
Rocket airplanes	140
Miniature fighters	144
Mistel - combination aircraft	145
Datafiles	154
Bibliography	163

INTRODUCTION

Powered flight has existed for over a hundred years. Throughout this whole century, aviation experienced a rapid and notable development. Initially viewed as toys for eccentrics, airplanes quickly turned into useful military tools and later into a vital means of transporting passengers and cargo.

Technological development has set the pace of aviation's evolution for the last hundred years. Materials and construction techniques, propulsion, avionics and systems, were all decisive for the evolution of aircraft. Research into aerodynamics was especially important.

Aviation evolved at an uneven rate. The speed of development was always influenced by technological advances in related areas. A very good example has to do with propulsion. For over four decades, the piston engine represented the sole form of power available to aircraft. By the late 1940's, the introduction of new engines like turbojets and turboprops, revolutionized aeronautics.

This Fiat CR 42 *Falco* preserved at the RAF Museum represents the last generation of military biplanes to see action during WWII. Photo: Claudio Farias.

The evolution progressed at different rates in civilian and military areas. During periods of peace it was usually in commercial aviation that the latest developments took place while investments in military technologies tended to diminish. In the years between the two world wars, many airliners had performances at least equal to or better than contemporary military aircraft.

In the first months of WWII, it was still possible to find in the military inventory of the warring nations the same design and construction techniques as had been employed at the end of the First World War. Aircraft such as the Gloster *Gladiator* (UK), the Fiat CR-42 (Italy) and the Heinkel He-51 (Germany) were all biplanes with steel tube and wooden structures, covered with fabric, and all had fixed landing gear. These were the same methods of design and construction as employed at the end of the First World War.

At the time these obsolete military aircraft were in service, it was possible for the common citizen to fly in aircraft such as the Douglas DC-3, with metal monocoque construction, streamlined monoplane design, and retractable undercarriage.

On the other hand, it cannot be denied that many innovations in aviation came from research initiated during wartime. Technology initially developed for military use was often adopted for civil use. Radar is an outstanding example. The two world wars, especially WW2, were accelerators and catalysts of aeronautical evolution.

The Heinkel He 178 was the first jet aircraft. It flew for the first time in August 1939 and is the true ancestor of many projects described in this book. Photo: courtesy USAF.

The most evident characteristic of the 1939-45 conflict was its total mechanization and the development of military doctrines based on this principle. It was a war between the world's technological and industrial powers.

In a conflict of this kind, technology assumes a decisive importance since there is an apparent equilibrium between the nations involved. Considering that the USA, England and Germany showed the same level of development, any small advance, even the smallest one in terms of the technological race, would almost certainly lead to great military gains.

That explains only some of the factors that set in motion a competition for the development of new advanced military technologies during the war. Technology grows in quantitative and also in qualitative terms, which means that technology advances by great leaps in a relatively small period of time.

The impressive rush caused by technological development during 1939-45, including in aviation, can also be explained by need to recover "wasted time" during periods of peace.

It must be noted however, that technology, even in a mechanized war between industrial powers, is not the only decisive factor. The USA's industrial might in conjunction with its enormous technical creativity, and the immense Soviet Armies, were the factors that decided the war in favor of the Allies.

Even if a nation possesses very advanced technology, if it is not capable of transforming it into high quality products manufactured on a massive scale, in order to make up for the huge losses that come from the use of these war machines, then this nation will be condemned to lose this war. This seemed to have been the fate of Germany.

Some authors claimed that Germany won this technological race. This seems to be especially true when German aeronautical research during the war is taken into consideration.

The apparent superiority of German scientists and technicians must be seen in the context of Germany's inferiority in terms of industrial capacity. There was no alternative but to invest in technology, developing extremely modern high performance war machines. This

combustion chamber

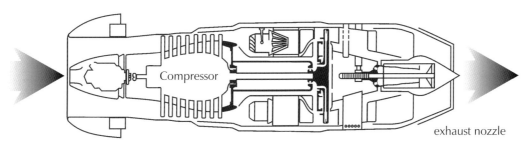

exhaust nozzle

The Junkers *Jumo* 004 turbojet was used in most of the German jets, including the Me 262, the Ar 234 and the Ju 287.
Drawing: Daniel Uhr.

strategy was never fully successful. It only served to postpone the inevitable while extending and maximizing the suffering of the civilian population.

It is also necessary to criticize the position taken by some authors who claim that if, for example, the Messerschmitt Me 262 had been put in production earlier, the conclusion of the war might have been different. Truth is much less dramatic: Germany did not have the materials, the work force, or the industrial plants to produce this famous fighter in the same quantity and with the same quality standards of aircraft such as the American P-51 *Mustang.*

However, one can not ignore the astonishing advances Germany made in aeronautical technology during the war.

It is possible to identify certain areas in which Germany was at the forefront of aircraft design. One notable example being jet propulsion (turbojets, rockets or pulse-jets) employed in military aircraft, combined with a series of revolutionary aerodynamic planforms, such as swept wings.

This lead is particularly evident in the amount of aerodynamic research which took place at many research institutes in Germany. The findings were promptly applied in many aircraft that managed to reach operational or prototype status, like the Messerschmitts Me 262 and 163 *Komet,* and the Horten Ho 229.

Little secrecy remains now but much has been written in the specialized press about the so-called "secret projects of the Luftwaffe," a term that is self contradictory. These projects have been deeply scrutinized in numerous publications over the last few decades.

The mystery that surrounded these designs began to dissolve in the last days of the war. As Allied troops entered Germany, the Soviet army from the east and especially the American and British from the west, the troops were often accompanied by intelligence teams whose mission was to locate, identify and capture as many aircraft, documents, and German scientists as they could.

It was of the utmost importance for the Western Allies and the Soviets, to take possession of as much German equipment and specialist knowledge as they could. Allied intelligence services had reasonably detailed information about the state of aeronautical engineering in Germany, from which they concluded that Germany was way ahead of the allies in many aspects of aircraft design. The race to obtain this information consisted of episodes that might fit perfectly in a James Bond movie, and was motivated by the increasing suspicion and animosity between the Western Allies and the Soviets. The Cold War started even before the end of WWII.

It was a great irony that the nation which least benefited from wartime German aeronautical research was Germany itself. Many of the projects featured in this book did not reach the hardware stage and were only put into practice after the war by the Allied nations. There will be a heated debate when it comes to discerning who was the greatest beneficiary.

At the end of the war, the United States acted to obtain as many as possible of the latest enemy aircraft and documents, and to secure the cooperation of German pilots and designers in order to better evaluate these machines. Under codenames such as "Lusty" (Luftwaffe secret technology), groups of military men and scientists explored recently occupied territory searching for Germany's technological treasures. Similar groups were formed in Great Britain and the Soviet Union, the latter capturing most of the German plants and installations in the eastern zone.

The Americans were also responsible for Project "Paperclip," whose intent was to transfer to the USA, the crème de la crème of aerospace scientists and technicians, and put them to work for the American government. Initially they were employed at Wright Field, and later at private research institutions. The most famous example is the rocket pioneer Werner von Braun, designer of the V2 weapon and later for the design of the Saturn V rocket that took the first astronauts to the moon.

Von Braun was only one of many. A great number of scientists and specialists in all branches of aeronautical research were brought to work in the USA, decisively contributing to America's technological leap after 1945.

A small list reveals the dimensions of the contribution these men made to American aviation:

Woldemar Voigt, designer from the Messerschmitt company, responsible for the Me 262, went to work at the Martin Aviation Company.

Dr. Richard Vogt, chief designer at Blohm und Voss, responsible for some of the more unusual projects illustrated in this book, worked at the Air Physics Corporation and as a consultant for Boeing.

Hans Multhop, aerodynamicist at Focke Wulf, creator of the "T" tail adopted in the Ta 183 project, went to the Martin Aviation Company.

Alexander Lippisch, the delta wing pioneer and designer of the Me 163 Komet, worked as a consultant for Convair.

Dr. Hans von Ohain was the creator of Germany's first turbojet. After working at Wright Field, he started an academic career at the Ohio State University, in Dayton.

Dr. Anselm Franz, who designed the Jumo jet engines, became vice-president of the Lycoming engine company.

Perhaps the USSR was the country that most benefited from the technological heritage captured at the end of the war. A huge collection of installations and aviation related documents, including the Junkers company headquarters in Dessau and the Ministry of Aviation in Berlin, fell to the USSR. Especially important was the capture of BMW's jet engine factory. A great number of Jumo 004 jet engines were taken. The first generation of Russian jet prototypes (MiG 9 and Yak 15) were powered by copies of German engines.

This does not mean that the Soviets lacked competent aircraft designers, on the contrary. Russia always counted on notable engineers such as Andrei Tupolev, Serguei Ilyushin and Arten Mikoyan. At the end of the war, some Russian aircraft had better performance than their German counterparts. The shortfall was more evident in the area of propulsion and in some aspects of aerodynamic research. Russia's technological deficiencies were caused by factors that had nothing to do with any supposed creative inferiority.

Just as the Americans had done before, the Russians also considered the possibility of sending German specialists to work in the Soviet Union.

The North American P-51 *Mustang* represents the pinnacle of piston engine fighter aircraft technology. This aircraft is preserved at the Imperial War Museum in London. Photo: Claudio Farias.

The Focke-Wulf Fw 190, seen here at the Imperial War Museum, was one of Germany's most versatile fighters, employed in many different roles. However Focke-Wulf never manufactured a jet aircraft during the war.
Photo: Claudio Farias.

Initially, Junkers personnel were instructed to continue in Germany. The development of many projects interested the Soviets, including the negative swept wing jet bomber Junkers Ju 287 and the pulse-jet ground attack aircraft, Junkers EF-126.

For security reasons, before long the Russians decided to transfer the German specialists to Russia. They created two OKBs (development bureaus). OKB-1, led by engineer Brunolf Baade, took charge of the development of Junkers aircraft, and OKB-2, led by Hans Rössing, resumed the development of the experimental supersonic aircraft DFS 346. These specialists remained in Russia up to the late 1940's, developing projects that had been initiated in Germany. They also designed new projects, 100% Soviet in origin.

There were instances in which the Soviets produced aircraft that were almost copies of foreign hardware (the most notable example was the transformation of the Boeing B-29 into the Tupolev Tu-4).

Another was the MiG I–270, a rocket research aircraft with a strong resemblance to the Messerschmitt Me 263/Junkers Ju 248. The Me 263 was a further development of the Me 163 *Komet* embodying many modifications in order to remove the many limitations of the original rocket interceptor.

The MiG had a retractable landing gear, a longer and more slender fuselage, and a dual chamber rocket engine. These improvements would solve the problem of retrieving the fighter after it landed and provide it with longer endurance and range.

Although the design was initiated at Messerschmitt, it was transferred to Junkers (hence the designation Ju 248), who built the prototypes that were discovered by Russian troops when they occupied Dessau.

A single complete prototype of the Ju 248 was examined by many Soviet design bureaus and in December of 1946, MiG presented its version of a rocket interceptor, the I-270.

At that time knowledge about the behavior of swept wings at high speeds was still slight in Soviet Union, so MiG opted for conventional straight wings. They also decided not to make the I-270 tailless, incorporating a "T" tail like the Focke-Wulf Ta 183 (whose plans were captured by Soviet troops – see Chapter 2).

Only two prototypes MiG I-270s were completed before the project was abandoned in 1948.

The presence and heritage of German specialists was felt even in South America. Focke-Wulf's chief designer, Professor Kurt Tank, established himself in Argentina. There he continued the development, albeit with many modifications, of the Ta 183 project which was transformed into the unlucky *Pulqui II*.

Reimar Horten also went to Argentina where he developed and built many sailplanes. He designed the IAe 37 delta wing research glider and also built a four engined flying wing transport aircraft, the IAe 38.

Professor Henrich Focke chose Brazil as the place to continue his work with VTOL aircraft.

In 1951, Focke and Brazilian engineers of the CTA (*Centro Técnico de Aeronáutica* – Aeronautics Technical Center, today known as *Comando-Geral de Tecnologia Aeroespacial*), developed the project *Convertiplano,* an aircraft similar in configuration and operation to Bell's Osprey V-22. The *Convertiplano* could take-off vertically, as a helicopter, and change its flight regime, through a mechanism that tilted the propellers, to achieve horizontal flight. This curious project used parts of a *Spitfire,* a radial engine mounted inside the fuselage, and a set of four tilting propellers. It was similar, at least in part, to the Focke-Achgelis Fa 269 fighter project.

This book documents a series of German aeronautical projects that incorporated extremely original ideas. It also demonstrates the connection between these projects and many of the aeronautical concepts put to use after 1945. The text identifies not only the areas in which this contribution was decisive, but shows also that some of these ideas proved to be aeronautical dead ends.

Many configurations and design solutions which are taken for granted now had their origins in aerospace projects of the 1940's.

Of course the war was won by real airplanes and not "paper planes" which never left the drawing board. The Allies won

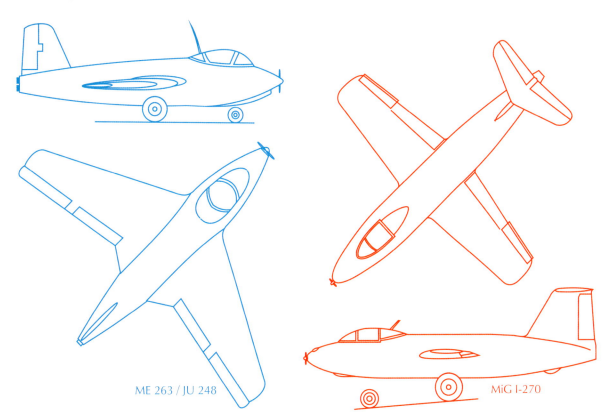

ME 263 / JU 248

MiG I-270

the war with aircraft of conventional design, such as the P-51 *Mustang*, the *Spitfire* and the Boeing B-29, which represented the peak of those technologies.

It is the contrast between the "between the g B- and production aircraft that makes the German high-tech projects so fascinating.

Some of the German designs were years ahead of their time, in fact so far ahead that they failed to contribute to Germany's war effort. Nonetheless, their contribution to aviation progress after 1945 is unquestionable.

Throughout aviation history, it was not always the *originators* who most benefited from their inventions. In many instances, the major beneficiary of great inventions, were those capable of perfecting and turning them into reality.

This book is divided into four chapters, each of them dedicated to one aspect of aircraft design, and each chapter is divided into many sub-chapters.

The Aerodynamics chapter analyses the different configurations of wings proposed and tested by German designers, including such variations as: swept wings, variable geometry wings and delta wings.

The chapter on Configurations refers to the different layouts adopted by the fuselage/wings/empennage unit that were employed in German projects. Tailless aircraft, flying wings, canards, asymmetric aircraft, and the "T" tail are discussed.

The VTOL chapter is concerned with aircraft designed to take off and land vertically. It also deals with rotating wing i.e. helicopter and convertiplane aircraft.

Finally, the chapter on Special Missions includes a series of aircraft designs that were conceived to accomplish very specific missions, such as: parasite aircraft, reconnaissance aircraft, mini-fighters, and rocket planes.

At the end of each chapter the specifications for the more representative projects are presented.

Aerodynamics

Professor Lippisch's "flying triangle", the DM-1 seated on its tail. The aircraft would assume a similar flight attitude while landing.
Photo: courtesy Peter F. Selinger.

This chapter analyzes the different wing shapes proposed for German aircraft projects during World War II. In all aspects of aeronautical design, it was in new wings that Germany excelled during the conflict. Swept wings, delta and variable geometry wings, all most important now, were by 1944 already on the drawing boards of manufacturers such as Arado, Heinkel, and Messerschmitt.

Aerodynamic research in Germany was directly linked to developments in propulsion, especially the turbojet but also rockets, pulse-jets, and ramjets. The increased thrust from reaction engines offered the possibility of reaching enormous speeds relative to aircraft powered by piston engines. They stimulated the development of new aerodynamic configurations in order to achieve these new performance levels.

Many of these configurations had also been studied by engineers in other nations like Boris Tcheranovski, who was responsible for a series of delta wing aircraft designs in Russia.

POSITIVELY SWEPT WINGS

The origin of the swept wing was in the mid-1930's. The concept was presented to the international aeronautical community for the first time in Rome. At the 1935 Volta Congress on High Speed in Aviation, German aerodynamicist Adolf Busemann lectured on *Aerodynamic Sustention at Supersonic Speed*. A member of the *Luftfahrtforschungsanstalt* (*LFA, Aeronautical Research Institute*), Busemann demonstrated with complex calculations how sweep could diminish aerodynamic drag at speeds above that of sound.

At a time when the maximum airspeeds of combat aircraft varied around 400 km/h, the lecture did not receive much international attention. Despite this, German aeronautical research institutes continued Busemann's research and he, together with Albert Betz, director of the *Aerodynamische Versuchsanstalt* (*AVA, Institute of Experimental Aerodynamics*) in Göttingen, registered the confidential patent Nr. 732/42 under the heading *Aircraft for Speeds Near the Speed of Sound*.

Soon swept wings moved beyond academic research. They became the foundation for secret designs by all the German manufacturers who quickly understood the benefits gained by the application of positive sweep (wings swept backwards) to aircraft designed to operate near the speed of sound. Projects such as the Ar E.560 from Arado, the Bv P 211/01 from Blohm und Voss,

The photograph shows the third prototype of the Me 262 being prepared for a test flight in 1942. Photo: courtesy Franz Selinger.

the research airplane DFS 346, the 1000x1000x1000 bomber from Focke Wulf, and the Heinkel He 162C are just a few among many designs using this configuration.

The Messerschmitt Me 262 had swept wings. Often misunderstood by later commentators, this was not intended to achieve higher speeds. It was determined by a need to adjust the balance of the aircraft after the adoption of engines which were heavier than those anticipated in the original design. Sweeping back the external sections of the wings corrected the balance. The aerodynamic gain was an unexpected bonus.

It was thereafter at Messerschmitt AG that the concept of swept wings found rapid acceptance, and by the end of the war a great number of designs using this solution had been outlined. The company produced a series of studies into high speed, swept wing versions of the Me 262 with high performance engines (Me 262 HG). More studies followed. These included designs such as the P-1106 and P-1110 (fighters), and the P-1107 and P-1108 (bombers) but none reached the prototype stage.

The first generation of American jets, the Bell XP-83 *Airacomet*, the Lockheed XP-80 *Shooting Star*, and the McDonnell FH-1 *Phantom*, were similar aerodynamically to the piston powered aircraft already in use by the American Air Force. Such cautious design approach was justified by the fact that the United States was entering virtually new territory. Experience there with jet designs was in its infancy compared with Germany or even Britain.

When the new aircraft began to approach the speed of sound, ways had to be discovered to diminish the aerodynamic drag for the next generation. Aerodynamicist Robert T. Jones, from NACA (National Advisory Committee of Aeronautics) had previously

The Lockheed F-80 *Shooting Star* preserved at the Museu Aeroespacial in Rio de Janeiro.
Photo: Daniel Uhr.

written a paper on swept wings that was not published because, it was argued, it lacked experimental evidence. Yet huge volumes of data had been generated in Germany. Such information, when it was found, was to be of inestimable value for the development of two of the most important post-war American combat aircraft: the North American F-86 *Sabre* fighter and the Boeing B-47 *Stratojet* bomber.

In the final days of WWII, the Advisory Scientific Group of the American Army, commanded by the aerodynamicist Theodore von Karman, headed for Europe in search of German aeronautical secrets. The group included not only Armed Forces personnel, but also engineers from the major aircraft manufacturers in the USA.

In 1944, the North American Aviation Company began the development of a carrier-based fighter for the Navy. The FJ-1 *Fury* was of conventional aerodynamic design. A land based version was offered to the Air Force, at first with straight wings like its Naval equivalent. This would eventually be transformed into the F-86 *Sabre*. Its design coincided with access to swept wing data from Germany. The wings of a Me 262 had been sent to North American and its leading edge slats were carefully studied. North American's designers were particularly interested in the benefits of wing sweep at speeds close to Mach 0.9. They obtained reports on the high speed versions of the Me 262 HG, which had engines buried

The North American F-86 *Sabre*. Photo: courtesy Bernardo Malfitano.

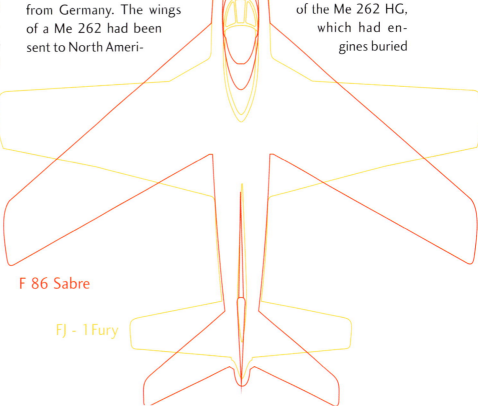

F 86 Sabre

FJ - 1 Fury

Luftwaffe - confidential - Fundamentals of modern aeronautical design

Boeing B-47's final design configuration owes a lot to data retrieved from Germany after the end of the war. Photo: courtesy NASA.

inside the wings close to the fuselage as well as sweep back.

The concept seemed promising and in August 1945 the Company decided to explore the possibilities. A program of wind tunnel tests was authorized and a 1:4 scale model of the swept wing was built in two weeks. When mounted on the F-86, this raised the maximum speed to 1120 km/h. After about a thousand separate tests, North American approved the project.

The wing shape adopted for the F-86 was virtually identical to the one considered for the Me 262HG. As a consequence, the *Sabre* was transformed at a decisive moment of its development from straight to swept wings. It was an extremely good decision. If the *Sabre* had kept its straight wings, its performance could have been inferior to the Republic F-84 *Thunderjet* and perhaps the history of the Korean War would have been different.

While the North American Aviation team was developing its new fighter, a team of engineers at Boeing confronted an equally difficult challenge: to develop a jet bomber. As with fighters, the new engines at first were simply fitted to aircraft such as the North American B-45 *Tornado,* the Convair XB-46, and the Martin XB-48 whose designs added little in terms of aerodynamic improvement compared with existing types such as the Boeing B-29 or the Martin B-26 *Marauder.*

Experiencing difficulty getting the required performance using a conventional design, George Schairer, head of the Boeing Department of Aerodynamics, was part of the team sent with Theodore von Karman in search of aeronautical treasures in Germany. The data they found validated the pioneering studies of Robert Jones and offered the solution they needed. They interviewed Adolf Busemann, who had originally proposed the theory of swept wings in 1935. Schairer, on returning to the USA, recommended that the bomber project at Boeing should incorporate them. Thus was born the Boeing B-47 *Stratojet*. From it evolved a long and successful family of civil and military aircraft including the B-52 *Stratofortress*, the Boeing 707, and the 747 *Jumbo*.

FORWARD SWEPT WINGS

At the same time as the advantages of the swept back wings for high performance fighters were discovered, problems also became evident. Favourable for high speed flight, swept back wings do not yield ideal performance at low speeds. This may lead to very high landing and take off speeds. Today such negative characteristics are dealt with on airliners by the use of high lift devices such as leading edge slats and double slotted flaps.

The installation of such complex equipment was not a viable alternative in aircraft designed for simplicity in manufacture and ease of operation in combat. German scientists considered ways to reconcile swept back wings with the necessity of better performance during take-off and landing. Junkers engineer Philipp von Doepp initiated a series of experiments with forward sweep. The results confirmed that forward sweep guaranteed good performance at both high and low speeds, although there was also a dangerous tendency of the wings to flex during turns. It was thought that either stiffening the wing structure, or improved positioning of the engines on the wings, were alternative ways to fight this aeroelastic effect.

Junkers used this unusual configuration in a series of bomber projects, beginning with the EF-122 that evolved to the Junkers Ju 287. To evaluate the performance of forward sweep at low speeds, Junkers built two prototypes of the Ju 287 using parts of other aircraft. The first of these flew on August 8th, 1944. A third prototype, configured as a production aircraft, was under construction at the end of the war. This was captured by Russian troops. The

Artist's impression of the Junkers Ju 287- production version.
Illustration: Daniel Uhr.

The experimental prototype Sukhoi Su-47 *Berkut* is the most recent application of the negative swept wing concept to a combat aircraft. Photo: courtesy Leonid Faerberg.

plane, technicians, engineers (headed by Brunolf Baade) and equipments (even including furniture) were promptly dispatched to the Soviet Union. Under the supervision of the authorities, a new prototype (designated EF-131) was finished using six BMW turbojets in clusters of three under the leading edge of each wing. It made its first flight in May 1947 and remained under evaluation until 1948.

From this German prototype, the Soviets developed a series of advanced variations (designated EF-140 and EF-140R), to fulfill bomber and/or reconnaissance missions. The Junkers heritage is clearly evident in both. Little of the original Ju 287 design remained in the last project created in Russia by these German engineers, the Model 150. This flew for the first time in September 1952, and on returning to East Germany, Baade initiated the design of an airliner based on it. Known as the "152," it flew for the first time in December 1958, but only two prototypes were built.

The Ju 287's configuration seemed sufficiently promising for other manufacturers to flirt with forward sweep. This was the case with the designs for the P-209/02 from Blohm und Voss, the surprising 1942 Focke-Wulf jet fighter project, and the swept forward version of the Heinkel He 162.

The forward swept wing concept gained little acceptance in the years immediately after the war. The inherent problems of aeroelasticity were difficult to overcome and demanded reinforced wings. This meant a serious increase in weight which eliminated the possible advantages.

One notable exception was the Convair XA-44/XB-53, designed about 1948. The project was originally conceived as an attack aircraft, formidably armed with twenty 0.50-caliber machine guns. Designed just after WWII, the XA-44 was one of the world's most advanced jet aircraft. It not only featured wings swept forward at $33°$, but it was also tailless.

Ad showing the passenger cabin of the Baade Type 152. Published in the Flieger-Jahrbuch 1959 (Verlag die Wirtschaft, Berlin).

Two prototypes were ordered in 1948 but its mission was changed to that of a fast medium bomber (XB-53). It would be powered by three General Electric J35 turbojets, would be able to carry twelve 1000 lb bombs, and fly at 938 km/h, a very impressive performance at the time. The project was cancelled so no prototypes were completed. Had it been built, the Convair bomber would surely have been a remarkable sight.

During the 1960's, engineer Hans Wocke, one of the Ju 287's designers, designed the 320 HFB *Hansa Jet* for the Hamburger *Flugzeugbau* (Formerly Blohm und Voss.).

The *Hansa Jet* was an executive aircraft in the same class as the *Learjet*, carrying up to twelve passengers. Its most evident characteristic was the swept forward wing.

More recently, with the advent of revolutionary composite materials such as carbon fiber and the increasing use of titanium in aircraft construction, it has been possible to create very light wings with extreme structural strength and stiffness. The concept of forward sweep therefore returned to

This nice photo shows the unique design of the HFB 320 *Hansa Jet*, the only corporate jet that employed forward swept wings. Photo: HFB. Rainer Niedrée's archive.

The experimental aircraft Grumman X-29 employed forward swept wings and also *canards*. Photo: courtesy NASA.

the drawing board as a way to create aircraft of extreme agility. Developed by Grumman in the United States, the experimental prototype X-29A went through a series of successful tests, beginning in 1984.

In Russia the OKB Sukhoi created the impressive Su-47 *Berkut*, an advanced experimental fighter. Although to the layperson its futuristic form may recall something from George Lucas' "Star Wars" saga, scholars of aviation history can surely identify a certain Germanic flair in this design.

CRESCENT WINGS

Another form of wing sweep explored by German designers was the so-called crescent wing. In

Photomontage showing a model of the extremely advanced Convair XA-44/XB-53. Photo: courtesy USAF.

this arrangement the wing possess different degrees of sweep along its length, increasing from the tip towards the root at the fuselage. The crescent wing was invented by engineers Rüdiger Kosin and Walter Lehmann of the Arado company and was patented in 1942 under number 844723. Kosin decided to test the idea in a series of prototypes for the Arado Ar 234.

The Handley Page *Victor*. Photo: courtesy Mike Freer.

Wind tunnel tests took place in the DVL and a decision was made to install one of the so called *Versuchsflügel II* (experimental wing II) on the Arado Ar 234 V16 prototype. It was not completed.

After the end of the war Reginald Stafford, the chief designer of the British company Handley Page, also began researching crescent wings. He had taken part in the British Fedden Mission, seeking Germany's aeronautical secrets.

The data captured at Arado impressed Stafford and he decided to incorporate crescent wings in the design of the Handley Page HP-80, later known as the *Victor*. In 1951 the crescent wing was tested in a British experimental aircraft, the Handley Page HP-88. This, a modified Supermarine *Swift*, served as a flying labora-tory. The *Victor* prototype flew for the first time in December 1952 and served the RAF for more than three decades.

VARIABLE SWEEP WINGS

Swept wings were a completely unexplored territory and scientists in Germany needed to discover their secrets very quickly. The existing high speed wind tunnels in Germany could provide only part of the required data necessary for the new jet aircraft, which were desperately needed by the Luftwaffe.

One of the questions requiring solution was the determination of the ideal wing sweep angle. From this came the idea, initially as a research tool, of a wing with variable angles of sweep. It was

The experimental Bell X-5. The multiple exposure photo shows minimum and maximum wing sweep angles. Photo: courtesy NASA.

The PANAVIA *Tornado* fighter-bomber. Photo: courtesy Stevie Beats

later discovered that this type of wing could solve problems found in aircraft with either forward or back sweep.

If it proved possible to build a mechanism that could modify the sweep angle of the wings in flight, the airplane could take off with wings unswept. Once in the air they could be inclined backwards for maximum efficiency at high speeds. When returning to base, the wings could once again be placed in the normal position, allowing lower speeds and a safer landing.

Messerschmitt initiated the development of an experimental aircraft to evaluate the ideal arrangement for future combat aircraft. The sweep, at whatever angle was required for testing, was to be set on the ground but not altered in flight. The project, known as the P-1101, went through various stages, from a research and experimental airplane to a single-engine fighter. The prototype was almost finished by end of the war and was captured by the American troops.

The idea interested Robert Woods, chief designer at Bell Aircraft Corporation, the manufacturer of the P-39 *Airacobra* fighter. The unfinished P-1101 was brought to the United States. Initially, there were plans to finish it and use it to test a series of American jet engines, and to incorporate a mechanism that would allow changing the wing sweep in flight. Later it was concluded that it would not be feasible to transform the P-1101 into a useful flying laboratory. It was decided instead to design an entirely new aircraft based on the P-1101's arrangement. This airplane was the Bell X-5, which flew for the first time in June 1951.

Later, variable sweep wings were seen in a great variety of military aircraft, such as the General Dynamics F-111, the PANAVIA *Tornado*, and the MiG 23/27. Similar installation was considered also for the projected American supersonic airliner, the Boeing 2707 SST.

Luftwaffe - confidential - Fundamentals of modern aeronautical design

The image shows the scissors wings system installed in the experimental AD-1.
Photo: courtesy NASA.

OBLIQUE WINGS

The indefatigable chief designer of the Blohm und Voss company, Prof. Richard Vogt, presented an unusual alternative solution to the problem of providing aircraft with wings of variable geometry, avoiding heavy and complex mechanisms.

His idea consisted of installing a conventional wing on top of the fuselage and rotating it on a vertical axis. In this way the right wing could swing back while the left wing would turn forward. The concept was considered for the P-202, conceived in 1942. This was certainly one of the most bizarre designs to come from Vogt's drawing board.

As strange as this idea may seem, it was perfected by Messerschmitt in an even stranger version. The P-1109 design foresaw not one, but two sets of pivoting wings, creating an extremely odd jet biplane.

The concept of a pivoting wing is so unusual that not a single experimental prototype was built during the war and it would be understandable if the idea had been completely forgotten.

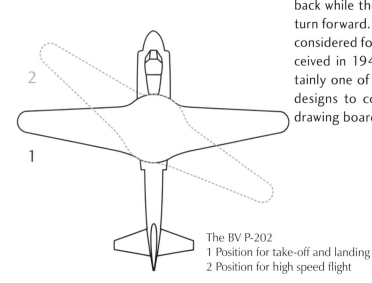

The BV P-202
1 Position for take-off and landing
2 Position for high speed flight

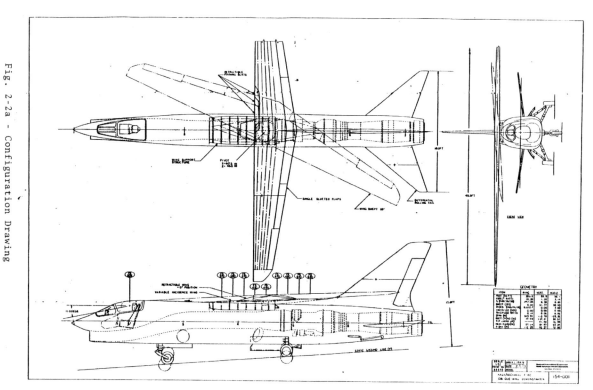

Fig. 2-2a - Configuration Drawing

The proposed supersonic F-8 *Crusader* Oblique Wing Research Aircraft.
Photo: courtesy NASA via Al Bowers.

Surprisingly, during the 1970's, the idea was resurrected in the United States by one of the most talented aerodynamicists of the 20th Century, Robert Jones, America's foremost enthusiast for oblique wings. In conjunction with NASA, he conceived a supersonic airliner using the so called *scissor* or *oblique* wing. Unfortunately this project did not go ahead. It would be very interesting to observe the reactions of the passengers when noticing that the wing turned above their heads.

Far from being a publicity stunt, the pivot wing idea was taken seriously and NASA sponsored the construction of an aircraft to test the concept in flight. This was the AD-1 and its design details were the creation of Burt Rutan. The experimental airplane had its initial flights in December 1979. The first complete wing rotation test (60^0) took place in April 1981, and thus validated Prof. Vogt's idea.

The success of the AD-1 prompted further investigation into the application of oblique wings to supersonic aircraft. Plans were on hand for the conversion of an F-8 *Crusader* to test the concept, but once again the prototype was never built.

Luftwaffe - confidential - Fundamentals of modern aeronautical design

Artist's impression of the Blohm und Voss P-188. Illustration: Daniel Uhr.

THE "W" WING

There was still one more way to obtain the benefits of variable sweep wings and at the same time avoid the weight and complexity of the mechanisms to alter the geometry in flight. It was called the "W" wing. The idea originated in Germany's most creative aeronautical company, Blohm und Voss. As the name implies, in this arrangement the inner sections of the wings are swept back while the outboard segments are swept forward, combining the best characteristics of both.

Once again Richard Vogt's imaginative genius produced another innovative solution to the problem of providing stability to jet aircraft during all flight regimes.

To make things even more interesting Vogt also included the possibility of altering the wing incidence to make it even more efficient during take off and landing. This radical design package would be applied to a family

The bicycle landing gear used in first generation American jet bombers like the Boeing B-47 and the XB-48 and XB-51, both designed and built by Martin, was initially tested in a converted B-26 *Marauder*. Photo: courtesy USAF.

of jet bombers identified by the design number P.188. In all there were four versions including both single and twin tails, single or double nacelles for the four Junkers Jumo 004 engines, and different fuselage shapes.

The P.188 family also exhibited another design solution that became popular after the war: the bicycle landing gear, as seen on the Boeing B-47 bomber.

Another swept wing planform appeared in some Focke-Wulf drawings. The original wing of the Fw 190 was replaced by one with positive sweep of the outer panels to cure compressibility effects, while the inner panels were swept forward. The curious resulting "M" shape enabled the same wing root attachment points as the conventional wing to be retained with the center of gravity at approximately the same position. This notion never advanced beyond very preliminary design stages.

AERODYNAMICS

37

Artist's impression showing the Fw 190 with an "M" wing. Illustration Daniel Uhr.

This artist's impression shows a hypothetical post war view of the Blohm und Voss BV 144 flying in the colors of *Sindicato Condor*, over Rio de Janeiro. Illustration: Daniel Uhr. Background photo: Eberius, Sindicato *Condor*, 1936.

This unusual wing shape found its way into some early Armstrong Whitworth and Bristol design studies that would lead, many years later, to the *Concorde*.

VARIABLE INCIDENCE WINGS

As mentioned in the previous section, Blohm und Voss designed a family of bombers with variable incidence wings. Although these never reached production, the concept was tried successfully in a prototype whose development had begun many years earlier.

In 1940, Deutsche Lufthansa requested Blohm und Voss to design a short and medium range, 18 seat, twin engined airliner, that would eventually replace its aging Ju 52 fleet.

The Bv 144, as it became known, was an advanced shoulder wing design. Sitting very low (75 cm) on its tricycle undercarriage, it was provided with slotted flaps and drooping ailerons to improve control during landing.

However, the Bv 144's most notable feature was the provision of wings whose incidence could be altered during flight. The tubular main spar (which was also used as a fuel tank) could be rotated up to $9°$. This would permit the aircraft to maintain a horizontal attitude during take off and landing, improving both passenger comfort and the pilot's view. After the fall of France when

The Vought F-8 *Crusader* displaying its variable incidence wing. Photo: courtesy NASA.

most of the country's aeronautical plants were put under German administration the Bv 144 project was transferred to Breguet (already involved in the manufacture of Fw 189 wings) who would finish the design and begin the construction of two prototypes. Only one of these was finished, and even then only after the liberation of France. Soon after "D-day," Allied investigation groups were already heading towards Paris. When the 18th Combined Intelligence Objectives Sub-Committee group visited the Breguet factory, they were immediately interested in the Bv 144 and recommended that one of the prototypes should be sent to the RAF in Farnborough, to be evaluated, but this never happened. The only prototype completed made three short flights in 1946. Soon the project was abandoned and the prototype scrapped.

While the shoulder wing arrangement, which was already quite common before the war, was adopted for aircraft like the Fokker F-27, the Handley Page *Dart Herald,* and the British Aerospace 146, the Bv 144's novel variable incidence wing was not used in airliners after the war,

In military aviation, the classic example of an application of the variable incidence wing twas the Vought F8 *Crusader*, which was an extremely potent and successful US Navy shipborne fighter used during the Vietnam War.

The image shows the Lippisch DM-1 and the team responsible for its construction: Herbert Dieks and Wolfgang Heinemann from the *Akaflieg* Darmstadt, and Klaus Metzner and Hermann Nenninger from the *Akaflieg* Munich. Photo: courtesy Hans-Peter Dabrowski.

The *Concorde* is seen here at *Tom Jobim* airport on the occasion of her last regular flight to Rio de Janeiro. Photo: Átila de Melo Coutinho (*in memorian*).

DELTA WINGS

Wings in a triangular, or delta, configuration were seen in the most modern combat airplanes during the 1950's, '60's, and '70's, the *Mirage* family, the British *Vulcan* bomber, the shipborne Douglas *Skyray* fighter, a series of Convair aircraft and the remarkable Lockheed *Blackbird* SR-71.

Triangular wings were also used in the *Concorde* airliner, considered one of the most beautiful flying objects ever created by man.

One of the greatest delta wing pioneers was Alexander Lippisch, who began his research with scale models in 1928, registering the patent number 615460. In 1929 he built a model glider called *Delta I* and in 1930, after success with the model, a manned version. This glider received an engine in 1931 and the *Delta* family evolved into a series culminating in 1935 with the *Delta IVa*.

Unknown to Lippisch, he had contemporary rival in Boris Ivanovitch Tcheranovski. This Russian dedicated himself to exploring the possibilities of both triangular and parabolic wings. Like his German counterpart, Tcheranovski built gliders and motorized aircraft, starting with his 1929 delta glider, the BITch-8. A rocket propelled experimental delta aircraft, the Kalinin K-15, was projected in Russia in 1936, but never built and in the United States in 1941 Russian immigrant Michael Gluhareff designed a piston engine fighter with triangular wings and pusher counter-rotating propellers. Although this did not go beyond wind tunnel model tests, its form was extremely advanced for the time.

Lippisch continued his work at DFS (*Deutsche Forschungsanstalt für Segelflug* - German Research Institute for Gliding Flight) developing tailless aircraft designs and creating further versions of the *Delta* series. After 1939 he was put in charge of the so called "Department L" at Messerschmitt and would participate in the initial development of the rocket propelled Messerchmitt *Komet*. Here Lippisch also outlined a series of more than 20 new types of aircraft. He also took part in the design of the *Komet's* subsequent variants: models B, C and the Me 263 fighter.

The relationship between Lippisch and Willy Messerschmitt had always been very difficult, culminating in Lippisch's departure from Messerschmitt in 1943 and his association with the Luftfahrtforschungsanstalt Wien (LFW). Here he once again pursued developments using delta wings, such as the P-11 and P-12.

Luftwaffe - confidential - Fundamentals of modern aeronautical design

This is how the Lippish DM-1 looked after extensive modifications that brought its configuration much more closer to that seen in the Convair XF-92. The aircraft was tested in the "Full Scale Wind Tunnel" at Langley Field.
Photo: Courtesy NASA via Hans-Peter Dabrowski.

The Convair XF-92 was America's first Delta jet plane. Photo: courtesy NASA.

The F-102 *Delta Dagger* seen in an ad published in National Geographic Magazine, August 1955.

Later, influenced by Professor Eugen Sänger's ramjet studies, Lippisch initiated the design of a supersonic fighter, the P-13a, as well as its glider test model the DM-1, which is analyzed in detail later in this book.

After the war, Lippisch embarked for the United States, under the auspices of the Paperclip Project. Wind tunnel tests with the modified DM-1 encouraged the American Army Air Forces to develop a combat aircraft using triangular wings. Convair received a contract to develop a research airplane, the XF-92 and Lippisch was brought in as a consultant. This tiny airplane initiated Convair's series of delta wing aircraft which included the F-102 *Delta Dagger* fighter and its perfected version, the F-106 *Delta Dart*, and the amazing XF2Y *Sea Dart* seaplane. The family reached its zenith with the B-58 *Hustler* bomber, which was capable of operating at Mach 2.

Luftwaffe - confidential - Fundamentals of modern aeronautical design

This Hispano HA 1112, manufactured in Spain and preserved at the *Museo del Aire in Cuatro Vientos* (next to Madrid) clearly displays its wing fences, just outboard of the wing guns. Photo: courtesy Rafael Garcia.

OTHER AERODYNAMICS INNOVATIONS - WING FENCES

Aerodynamic wing fence devices, developed in German laboratories during WWII, are often seen now on the swept wings of jet aircraft, both civil and military.

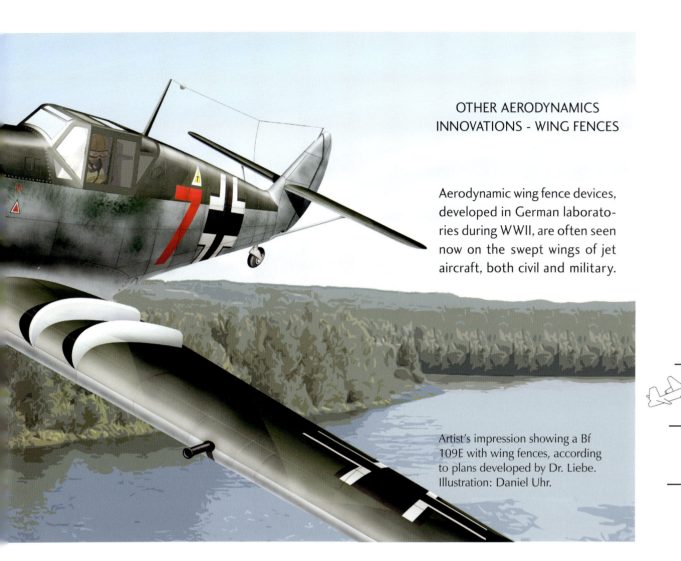

Artist's impression showing a Bf 109E with wing fences, according to plans developed by Dr. Liebe. Illustration: Daniel Uhr.

Wing fences were a common feature of many aircraft of the MiG family, including this MiG-15 preserved at the USAF Museum. Photo: courtesy USAF.

They control the airflow over the wings to prevent its span-wise movement that would lead to lift and control losses. They were first conceived by Dr. Wolfgang Liebe at the DVL, in Berlin.

The story began in 1938 when Messerschmitt was trying to improve the Bf 109B's stall behavior. As happened before with the Bf 108 *Taifun* (and many other aircraft), the Bf 109B was fitted with leading edge slats, but they did not solve the problem completely.

Liebe continued the tests with a Bf 109E in Augsburg. By gluing wool tuffs over the wing's surface he observed the airflow separation. After installing chordwise vertical metal barriers, called fences, over the wing, he obtained much better handling. This made him believe that the leading edge slats could be eliminated. Wing fences were not adopted then and remained secret until the end of the war.

Although no Bf 109 ever flew in combat with wing fences, the device made a comeback in the *Hecho en España* (Spanish) version of the 109, the *Hispano Aviacion* HA 1112K *Buchón* (Pigeon). These, based on Bf 109G airframes, were powered by Rolls-Royce *Merlin* engines and had 20mm wing-mounted cannons, just like their 109E ancestors. The wing fences were mounted just outboard of the guns. The Spanish retained the leading edge slats.

After WWII, wing fences became very popular on aircraft with extremely swept wings. They became almost standard in the MiG family of fighters. Fences were also found in commercial aviation, displayed prominently on the wings of the *Caravelle*, for instance.

Variable incidence wings

Blohm und Voss BV 144

The Blohm und Voss airliner was very similar in concept (with the exception of its variable incidence wing) to the Douglas DC-5, which was another shoulder wing airliner having a tricycle undercarriage, that flew for the first time in 1939. After securing the Lufthansa order, Blohm und Voss tested the variable incidence wing on a Bv 140 torpedo plane and results showed that the concept was sound and very promising. The Bv 144 owes its preliminary design to Hans Amtmann. Due to a lack of design capacity, a team of Breguet engineers, transferred from France to secluded offices at Blohm und Voss, in Hamburg, would be put in charge of detail design.

Passengers would be carried in two separate cabins. Each cabin would carry nine passengers in a 2 +1 arrangement, and a full scale cabin mock-up was built.

Construction took place in France and it seems that the only completed prototype was configured as a cargo plane instead of airliner, perhaps considering its future use by the Luftwaffe.

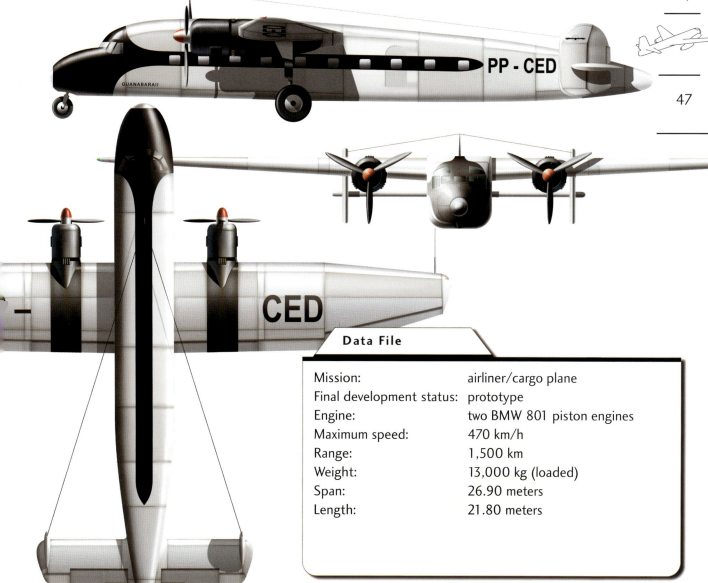

Data File

Mission:	airliner/cargo plane
Final development status:	prototype
Engine:	two BMW 801 piston engines
Maximum speed:	470 km/h
Range:	1,500 km
Weight:	13,000 kg (loaded)
Span:	26.90 meters
Length:	21.80 meters

Luftwaffe - confidential - Fundamentals of modern aeronautical design

Swept forward wings

He 162

The forward swept wing version of the He 162 would have been powered by the HeS 011 engine and would have a butterfly tail.

Data File

Mission:	single place day fighter
Final development status:	design
Engine:	one HeS 011 turbojet with 1,300 kg thrust
Maximum speed:	920 km/h
Range:	--------------
Weight:	3,650 kg (loaded)
Span:	8.00 meters
Length:	9.25 meters
Armament:	two Mk 108 cannons

Ju 287

Ju 287 - Four-engined bomber project

Data File

Mission:	Heavy bomber
Final development status:	prototype
Engine:	four Junkers Jumo 004 turbojets
Maximum speed:	680 km/h
Range:	1,500 km
Weight:	20,000 kg (loaded)
Span:	20.10 meters
Length:	18.28 meters
Armament:	-----------------

The 1943 Junkers heavy jet bomber, designated Ju 287 incorporated swept forward wings which had not been tested on operational military aircraft. The company decided to build a prototype to test its low speed behavior. To expedite construction, as many existing parts as possible were used. The fuselage came from a Heinkel He 177, the empennage from a Junkers Ju 188, main wheels from a Junkers Ju 252, and the twin nose landing gears from captured Consolidated B-24 *Liberator* bombers. Two prototypes were built using this ingenious component combination but both were destroyed.

Swept back wings

Me P-1110

The Me P-1110 project of January 1945 sought to establish the ideal arrangement for engine and other components installed inside the fuselage, to obtain maximum aerodynamic efficiency.

The reduction of the fuselage cross section area was one of the most important design parameters.

There were two versions: the first using wings very similar to those of project P-1101 (swept at 40°) and a conventional swept empennage.

The engine intakes were positioned flush (NACA style) on the fuselage sides. The 1946/46 Swedish SAAB *Lansen* fighter was very similar. The second variant of the P-1110 would have had an annular air intake around the fuselage and a butterfly tail.

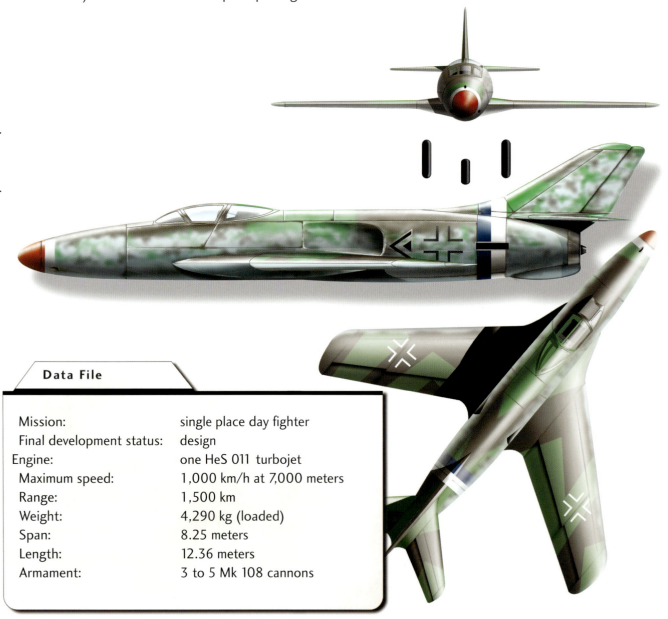

Data File

Mission:	single place day fighter
Final development status:	design
Engine:	one HeS 011 turbojet
Maximum speed:	1,000 km/h at 7,000 meters
Range:	1,500 km
Weight:	4,290 kg (loaded)
Span:	8.25 meters
Length:	12.36 meters
Armament:	3 to 5 Mk 108 cannons

Crescent wings

Ar 234 V16

The Arado company prepared four prototypes for the installation of laminar crescent wings: Ar 234 V16 (PH+SW), Ar 234 V18, Ar 234 V26, and the Ar 234 V30. The V16 prototype would be powered by two BMW 003R which combined conventional turbojets with a BMW 718 rocket engines to boost performance for short periods. Besides the crescent wings, these aircraft would also feature swept back horizontal stabilizers. Construction began only for the V16 which would have tested the *Versuchsflügel II* (experimental wing II) with a 37° sweep at the root and 23° at the tip.

Data File

Mission:	experimental aircraft
Final development status:	prototype
Engine:	two BMW 003 turbojets
Maximum speed:	-----------------
Range:	-----------------
Weight:	-----------------
Span:	13.20 meters
Length:	12.85 meters
Armament:	-----------------

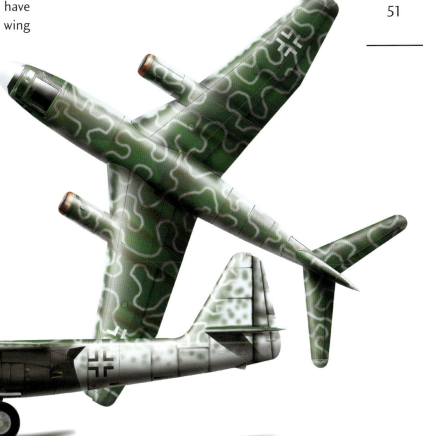

Variable sweep wings

Me P-1101

The P-1101 design number encompassed many different projects that eventually led to a 1944 proposal for an experimental aircraft to obtain data about the ideal wing sweep angle. A prototype was built with a mechanism allowing the angle to be set at three positions while on the ground: 35, 40, or 45 degrees. Even before the prototype was finished, its role was once again changed and it was decided to transform it into a multifunctional fighter. It would eventually be developed into day fighter, a reconnaissance aircraft, an interceptor armed with missiles (using either the Henschel Hs 298 or the Ruhrstahl X4), and there was even a variant adapted for night fighting. It was initially considered that the P-1101 could be an effective alternative to the Me 262, not only in terms of manufacturing costs but also in performance. The virtually completed prototype was captured and taken to the USA, together with its designer Woldemar Voigt. In America the aircraft was thoroughly studied at Wright Field and then offered to the Bell Aircraft Company who used it as a starting point for the development of the Bell X-5, an experimental aircraft that could alter its wing sweep in flight.

Data File

Mission:	single place day fighter
Final development status:	prototype
Engine:	one HeS 011 turbojet
Maximum speed:	985 km/h
Range:	1,500 km
Weight:	4,064 kg (loaded)
Span:	8.25 meters (at 40° wing sweep)
Length:	9.17 meters
Armament:	four Mk 108 cannons and 500 kg of bombs

Oblique wings

BV P-202

This project is one of Richard Vogt's most unusual creations. It would be a single seat fighter using the pivot wing. The mechanism would allow a 35° rotation. It is not known if rotation would be continuous or set at pre-determined intervals.

If the pilot would have had the possibility to choose any sweep angle between maximum and minimum values, at least in theory, he would have an extremely adaptable aircraft under his command. Flaps and landing gear could be lowered or raised only with the wing in the perpendicular position.

Data File

Mission:	single place day fighter
Final development status:	design
Engine:	two BMW 003 turbojets
Maximum speed:	877km/h at 3,500m
Range:	---------
Weight:	5,400 kg
Span:	10.06 meters (at 35° wing sweep)
	11.98 meters (at 0° wing sweep)
Length:	10.45 meters
Armament:	one Mk 108 cannon between the engine air intakes and two MG 151/20 cannons at the fuselage sides

Delta wings

Lippisch P-13a

The P-13a fighter is one of Alexander Lippisch's most fascinating projects, not only for its delta aerodynamic layout, but also for its power train: a ramjet which would use coal granules as fuel. The inside the intake tube, would contain 800kg of coal granules mixed with an oxidizer. Combustion would be initiated by gasoline flame.

Uniform burning of the fuel would be achieved

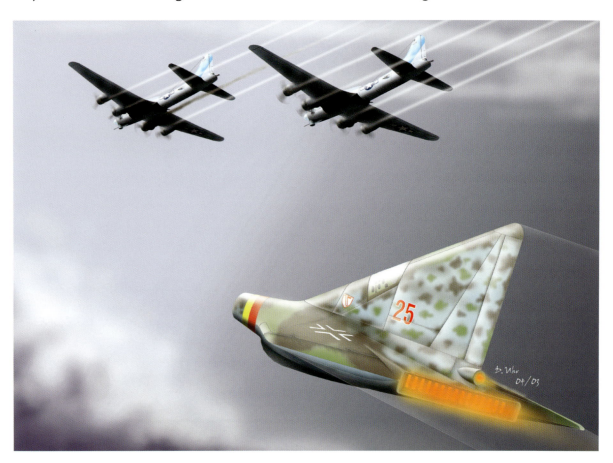

P-13a followed a long line of delta designs. Its origin can be traced back to the 1930 *Delta I* glider.

The P-13a wing would have a 60° sweep, with the cockpit at the base of its immense fin and rudder.

A Walter rocket engine would accelerate the fighter to a speed at which the ramjet would kick in.

The ramjet would be fed by a complex and ingenious fuel system. A circular wire basket, installed by rotating the wire basket at 60 rpm. This system had been tested by Dr. Eugen Sänger and it was believed that it would propel the P-13a to supersonic speeds. The project would progress in a series of steps, from scale models, full sized gliders and finally prototypes carried (in the *Mistel* arrangement) by other aircraft. Many gliding models were built and flown and a wind tunnel model was tested at the *Aerodynamischen Versuchanstalt* (AVA) in Göt-

Data File

Mission:	single place supersonic ramjet day fighter
Final development status:	wind tunnel models and glider prototype
Engine:	one ramjet plus a Walter booster rocket engine
Maximum speed:	1,600 km/h
Range:	150 km
Weight:	2,300 kg (estimated)
Span:	6.00 meters
Length:	6.70 meters
Armament:	two Mk 108 cannons

Lippisch DM-1

In order to evaluate the P13a's low speed characteristics in flight, it was decided to build a test glider, the Lippisch DM-1. This designation originated at Darmstadt University in the *Flugtechnische Fachgruppe* (*FFG*, generally known as *Akaflieg Darmstadt*), which was in charge of the construction of the full-scale glider.

The FFG was destroyed during a bombing attack in 1944, so the construction was transferred to Munich, hence the DM-1 designation. The glider was built of timber, plywood, and steel tubes. It was planned to mount it, *Mistel* fashion (see Chapter 4), above the fuselage of a twin engine Siebel 204A. When the correct altitude was reached, the DM-1 would be released and would return, gliding, to base.

Consideration was given to the installa- tion of solid fuel rockets under the fuselage which would allow the DM - 1 to reach speeds up to 800km/h. After the war, the glider was discovered by allied troops who decided to finish its construction. Plans were made to perform tests while still in Germany, and to this end, the DM-1 was to be mounted over a C-47 *Dakota*. The tests never took place.

Orders came to transfer the prototype to the USA, where it would be evaluated in the full-scale wind tunnel at Langley under the auspices of NACA. To everyone's surprise, considering its exotic appearance, wind tunnel tests showed only mediocre results.

An extensive series of modifications was implemented. Each small modification led to better results and in the final configuration, tests were consid- tingen. Results were considered satisfactory and it was decided to build an engineless full-scale model, to be launched from a carrier aircraft and then gliding freely to the ground. This would be the Lippisch DM-1, described in the next datafile.

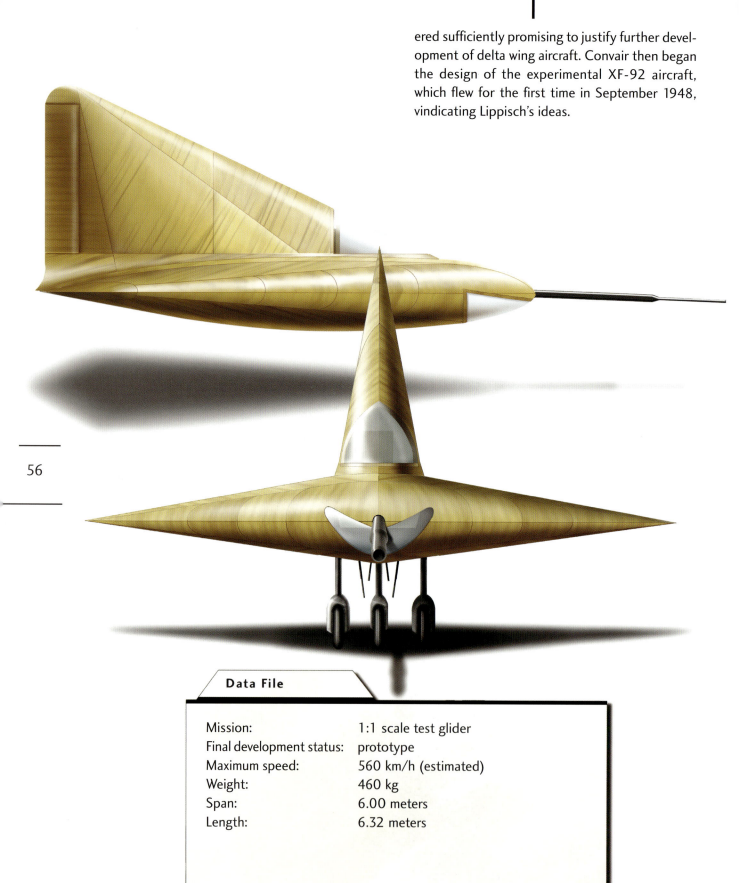

ered sufficiently promising to justify further development of delta wing aircraft. Convair then began the design of the experimental XF-92 aircraft, which flew for the first time in September 1948, vindicating Lippisch's ideas.

Data File

Mission:	1:1 scale test glider
Final development status:	prototype
Maximum speed:	560 km/h (estimated)
Weight:	460 kg
Span:	6.00 meters
Length:	6.32 meters

Configuration

This beautiful ad puts in evidence the unmistakable "T" tail of the Vickers VC-10, one of many airliners of the 1960's to adopt Hans Multhopp's solution. Photo: BOAC.

This chapter presents a small selection of the diverse configurations adopted by German production and prototype aircraft during the war.

Configuration can be defined as the way the elements that form an aircraft, normally wings, fuselage and tail, are arranged.

To the layperson, an airplane consists of a fuselage, roughly of cylindrical form; a pair of wings, generally located half way between the extremities, a tail at the rear formed by a vertical fin, rudder, a horizontal stabilizer and elevator. This is the most common configuration seen in civil aircraft.

However, with current military aircraft, there is a greater variety, reflecting the diversity of missions they must fulfill.

As seen in Chapter 1, contemporary military aircraft have adopted configurations whose evolution began during WWII, or even earlier.

Among these, tailless airplanes distinguish themselves. They can be divided into two sub-categories.

Those that still possess a visible fuselage and one or more vertical tails will be referred as tailless aircraft.

In the second sub-category, referred to hereafter as flying wings, designers have opted to eliminate both the tail and the fuselage, or blend it into the wings, thereby reducing the aircraft to a configuration approximating a pure wing. The American Northrop B-2 *Spirit* stealth bomber is the best example of an aircraft that uses this configuration today.

This chapter will also analyze some other arrangements, such as canards, 'T' and 'Butterfly' tails which offer alternatives to the orthodox empennage, and other unorthodox fuselage configurations such as twin fuselages and asymmetry.

TAILLESS AIRCRAFT

The partial elimination of the aircraft empennage has been the aeronautical designers' order of the day since the beginning of aviation. The search for aerodynamic efficiency has justified the removal of any superfluous element in order to achieve less aerodynamic drag and therefore better performance.

This approach preceded German WW II research by many years and was duplicated by almost every nation that endeavored to design aircraft.

As early as the 19th century, the Frenchman Alphonse Penaud proposed a very advanced design for the time, integrating the wings and the empennage into a single body.

As with other visionaries of his time, Penaud's design was never built, but it already represented a clear vision of design that would be developed many years later.

In 1907, the Austrian Igo Etrich built tailless glider aircraft, inspired by the seed of the *Zanonia macrocarpa*, a tree from Java whose seeds have the ability to glide.

These studies continued in England, through the pioneering work of designers such as John William Dunne and Professor Geoffrey T. Hill. They built and tested a successful family of tailless aircraft, but this achievement was not enough to guarantee the large scale use of this kind of aircraft, either for civil or military operators at that time.

Problems of stability and a skeptical attitude towards this kind of aircraft, postponed the acceptance of tailless airplanes until WWII.

As mentioned previously, the pressures imposed by the war opened the way for the adoption of unconventional designs as a way of obtaining performance gains that would place a country in a situation of superiority in relation to its enemies.

This was particularly evident in Germany. German designers carried out innumerable incursions into the territory of tailless aircraft, as well as into the studies of new aerodynamic configurations for the wings during the period 1933 to 1945.

The Horten IX V2 being prepared for takeoff in Oranienburg, in February 1945. Photo: courtesy Peter F. Selinger.

Luftwaffe - confidential - Fundamentals of modern aeronautical design

In fact in Germany, just as happened in Britain and Russia, since the beginning of the 20th century a solid research tradition in unconventional configurations also existed.

One of these pioneers was Professor Alexander Lippisch, mentioned in the previous chapter for his pioneering work with the delta wing.

In the early 1920s, even before he started work that would lead to project DM-1, Lippisch initiated another line of investigation that put him in the path of tailless aircraft.

This family of aircraft received the generic name *Storch* (stork) and culminated in the DFS 39, which can be considered a direct forerunner of the Me 163 *Komet* rocket fighter, as well as the research rocket airplane DFS 194, which is another ancestor of the *Komet*.

The history of the *Komet* has already been told in countless publications, but it seems fitting here to mention its extremely advanced character. It married an unorthodox aerodynamic configuration with an equally bold propulsion system, the rocket engine.

The successful tests of experimental aircraft before the war combined with the setbacks suffered by the Luftwaffe after 1940, seemed to indicate that the introduction of unconventional designs would be more easily accepted in the future generations of German combat aircraft.

No other German aeronautical company was so associated with the "unconventional" concept as Blohm und Voss with its chief designer Professor Richard Vogt. His fertile and creative mind gave birth to a torrent of projects that seemed to have originated in a sci-fi movie. Unfortunately this bold design approach failed to obtain commercial success in the same volume. Both Aviation Ministry officials and the Luftwaffe seemed to consider Vogt's creations as aeronautical freaks and remained attached to traditional design solutions.

Vogt detailed a plethora of projects using the tailless configuration, beginning with the P-208 family of fighters, still powered by piston engines, and quickly going to jet propelled designs, such as in the P-212 and P-215 fighter family.

Vogt acknowledged the problems of stability inherent to the tailless configuration and did

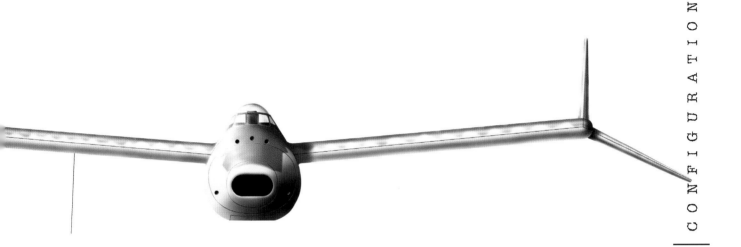

The Blohm und Voss P-215. This three place night fighter project was designed in January 1945. It was an enlarged version of the P-212 project. Thus it was possible to house the crew of pilot, radar operator and gunner and, at the same time, provide the necessary space for a large fuel load, the heavy armament and a complete state of the art avionics suite. The wings were swept at 30° and the armament would include a combination of large caliber guns firing forward and also a remote controlled turret firing aft. Illustration: Daniel Uhr.

CONFIGURATION

not completely eliminate the fuselage. Not only in the P-212 project, but also in the P-215, a voluminous fuselage is present, used to house engines, crew and heavy armament. This can be seen in the P-215 night fighter, armed with up to five 20mm cannons and operated by three crew members.

Vogt's designs tried to solve the stability problems in two ways: either by positioning small vertical rudders over the wings or by the adoption of down turned wing tips. In doing so Vogt tried to reconcile the best of two worlds: the aerodynamic efficiency of a tailless airplane with the stability provided by the empennage.

To arrive at this ideal configuration, Blohm und Voss undertook a complete program of wind tunnel tests, using scale models. Many configurations were tested before arriving at the definitive arrangement, which employed down turned wing extensions.

Blohm und Voss did not hold the monopoly on tailless aircraft projects. Both Messerschmitt and Junkers also considered this configuration for their advanced designs.

Professor Willy Messerschmitt's design bureau produced a good number of tailless aircraft designs. These ranged from cargo planes to a complete family of fighters, among them the P-1111 project.

This fighter almost achieved the pure flying wing layout as it tried to integrate wing and fuselage into a single body. Yet just like Richard Vogt at Blohm und Voss, in order to guarantee the stability of his aircraft at high speed, Messerschmitt decided not to eliminate the vertical rudder.

The Junkers company had a solid tradition in tailless aircraft design beginning in the first decades of the 20th century (see section FLYING WINGS). Junkers also developed jet propelled tailless designs, such as the EF-128 project.

The EF-128 design foresaw a series of applications, mainly as a single-place day fighter and as a twin-seat night fighter. It was not a pure flying wing as it included a short fuselage and a set of vertical rudders on the wings.

After the war, not only the Americans but also the British continued the research on tailless aircraft, using data obtained from the German aerodynamic research institute archives or directly from German manufacturers.

In the summer of 1945, occupation troops instructed Messerschmitt employees to prepare a series of reports and to restore documents related to the research and application of swept wings and tailless aircraft.

These reports, as well as German designers and scientists, had been placed at the disposal of a group of American and English engineers, which included R. G. Bishop and R. M. Clarkson, both de Havilland Company designers.

In England after the war, de Havilland, manufacturer of the formidable *Mosquito* fighter/bomber aircraft, was soon tasked by the government with the development of a high speed research aircraft. De Havilland chose to modify a Vampire fighter in order to obtain data about its behavior at transonic speeds.

The twin tail and wings of the Vampire were removed. A set of swept wings were substituted and in an audacious (and ultimately fatal) decision, it was decided to increase the angle of sweep of the wings from 23.3^0, as seen in the Me 163 *Komet*, to 43^0. The twin tail of the *Vampire* gave way to single large fin and rudder, also provided with sweep.

Designed by Bishop, the aircraft received the designation DH 108, commonly known as the *Swallow*. It made its first flight in May 1946, piloted by Geoffrey de Havilland.

After a series of low speed tests, it was decided to use the aircraft in an attempt to break the world speed record. The airplane was prepared and once again Geoffrey de Havilland was the pilot in charge.

Spectators witnessed the aircraft disintegrate in flight during the attempt, while reaching the speed of Mach 0.875. This accident demonstrated the risks faced when venturing into a virtually unknown flight regimen, in which aircraft behavior was still almost unpredictable.

Other prototypes of the DH 108 were built and one of them was the first British airplane to exceed the sound barrier (in an uncontrolled dive), in September 1948.

Unfortunately other tragedies would mark the airplane's career. Two other pilots would lose their lives piloting the DH 108, making the cost of obtaining such data very questionable.

In the USA, an aircraft of very similar configuration was built in order to fulfill the same type of investigation carried out by the DH 108.

In contrast to de Havilland which had little experience in the design of tailless aircraft, Northrop had long been involved in research and development of flying wing aircraft. It was the natural choice for the design of the experimental X-4 aircraft.

An extensive test program was undertaken by two prototypes, gathering a valuable volume of data which indicated that the configuration would not be ad-

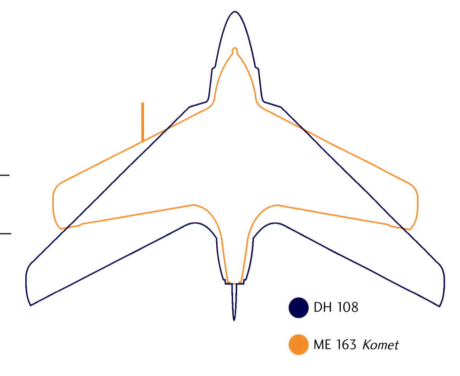

● DH 108

● ME 163 *Komet*

The experimental Northrop X-4. Photo: courtesy NASA.

equate for supersonic aircraft. In contrast to the British program, no serious accident marked the career of the X-4 and this airplane probably was designed without the aid of data obtained in Germany, although the decision for its construction (in 1946) could have been directly influenced by the impact of Me 163 *Komet's* combat debut.

FLYING WINGS

The dream of creating an aircraft whose configuration is reduced only to lift generating components (the wings) has filled aeronautical designers' minds since the beginnings of aviation history.

The logic behind this dream is very simple: eliminating any component that does not contribute to lift (fuselage, horizontal stabilizer and tail) will substantially reduce the aircraft's weight and improve its aerodynamic efficiency.

An aircraft designed under such principles will have greater autonomy, and will also be more fuel efficient thanks to its nearly perfect (at least theoretically) aerodynamics.

This is why the flying wing concept was pursued by designers all over the world as a sort of aeronautical engineering "Holy Grail."

No country can assume the exclusive paternity of this idea since many creative designers have tried to perfect this type of configuration over the last 100 years.

By 1910 in Germany, professor Hugo Junkers presented the patent of an aircraft that completely eliminated the tail and placed engines, passengers and load inside of a wing of great thickness. This aircraft was never built, but it is an important landmark on the road that would lead to the development of a viable flying wing.

In the following years, Junkers would create a series of designs that used this configuration, but they also failed to reach the prototype stage.

In the United States the most prominent enthusiast of the flying wings was Jack Northrop. His research began with a project

The Northrop B-2 *Stealth* bomber. Photo: courtesy USAF.

Luftwaffe - confidential - Fundamentals of modern aeronautical design

The Northrop XB-49 was the final development in Jack Northrop's quest for the perfect flying wing aircraft. The concept would be vindicated many years later in the form of the B-2 *stealth* Bomber. Photo: courtesy USAF.

that still used a conventional tail linked by two booms to a thick wing which housed the engine and the cockpit.

This first project was followed by a family of aircraft that could already be considered pure flying wings, culminating in the giant XB-35 piston engine bomber as well as the jet propelled model XB-49.

Northrop's efforts and tenacity finally achieved worldwide recognition with the production and introduction in service of the B-2 Stealth bomber, a direct descendant of the flying wings of the 1940's.

In Germany, the flying wing found two indefatigable supporters in brothers Walter and Reimar Horten.

They grew up in a country recently defeated in World War I, and where military forces were severely restricted by the terms of the Versailles Treaty.

It is a well known fact that Germany resorted to all kinds of subterfuge in order to circumvent the limits imposed by the Treaty of Versailles, including the establishment of airplane factories in other countries (as in the case of Dornier in Switzerland), or even the creation of secret bases in Soviet Union for the training of both Russian and German military pilots, long before Hitler and the Nazis came to power.

Germany's secret rearming plans included the creation of a modern and powerful air force (the future Luftwaffe), and actions were taken in order that a new generation of pilots would be ready when Hitler's expansionist plans were put in motion.

In the 1920's and early 1930's, the path to reach this goal lead to the creation of an "aeronautical spirit" among Germany's youth, and the tool used to achieve it was sport aviation.

Beginning with model making and quickly going on to sailplane flying (a practice initiated just after WWI, even before the Nazi Party grasped power), a whole generation of aviators were fully ready to become civil and military pilots when time came.

Sailplane construction and competitions were the instruments that allowed the embryonic creation of the most powerful air force of that time. It also allowed the origination of a new kind of weapon that would be decisive for the early victorious campaigns of Hitler's Lightning War (*Blitzkrieg*): the military combat glider, perfectly represented by the DFS 230.

The Horten brothers grew up among this dynamic environment. Their background was an eminently practical one, derived from many years of experimentation in model competitions and later by sailplane flight in the Rhön Mountains (Wasserkuppe), the Mecca for sailplane flight in Germany.

The brothers were inspired by the work of Alexander Lippisch. Yet using only their models which were conceived without the aid of a formal academic education or access to wind tunnel testing, the two youngsters developed the basics for lift calculation and even more importantly, finally the correct determination of the center of gravity in flying wing aircraft.

The Horten VII taxiing and preparing for takeoff. Photo: courtesy Peter F. Selinger.

Using a systematic approach, the Hortens developed a whole family of sailplanes and motor planes that allowed them to perfect the flying wing concept. By 1943, they had already amassed a volume of data that was to become the foundation base for their most ambitious project: a twin engine jet fighter using the all wing configuration (*Nurflügel*), or flying wing.

The Horten IX was developed using the same methodology the Hortens used in their pre-war machines. Initially a glider version was built, the Ho IX V1 and employed to discover its low speed behavior. Once the correctness of the design was confirmed, a fully motorized version would follow, powered by two Jumo 004 jet engines.

Future Ho IX pilots would be introduced to the unique flight characteristics of the flying wings, by learning to fly the Horten Ho VII, a piston engine flying wing designed by the Hortens in 1943.

This two-place aircraft was powered by two Argus As10 engines, thus becoming the ideal transition plane for future flying wing pilots.

In a later stage, the Ho IX design allowed for the construction of biplace versions,

Artist's impression of an operational Horten IX.
Illustration: Daniel Uhr.

The IAe 41 *Urubu (vulture)* preserved at the *Museo Aeronáutico in Buenos Aires* (Morón). Photo: Claudio Farias.

Gotha initiated the manufacture of new prototypes, but their executives saw in this contract the opportunity to offer an "improved" version of the Ho IX to the Luftwaffe.

It should be mentioned that Gotha had no previous experience in the design of flying wing aircraft and the designer in charge, engineer Rudolf Göthert, was the antithesis of the Horten brothers in terms of his background. Göthert was an academic whose knowledge came from laboratories and wind tunnels, but without any previous experience in the design of tailless aircraft.

The project that would compete with the Ho IX was the Gotha P-60, a compromise between a pure flying wing and a conventional aircraft.

specially suited for night fighting duties as well as advanced pilot training.

The first prototypes of the Ho IX were built in a car garage in Göttingen and used the same techniques and materials employed in their preceding machines. That is a central structure of welded steel tubes, covered with plywood, to which were attached the wings, which were entirely made of wood.

Tests with the first powered prototype, the Ho IX (V2), showed good flight characteristics, but it was destroyed attempting an emergency landing, after one of the engines flamed out.

The accident did not diminish Luftwaffe's interest in the flying wing concept and the Ho IX was ordered into production immediately, receiving the official designation Ho 229.

The Horten brothers did not have production facilities of their own, so it was decided that production machines would be contracted with the Gotha Company, better known for the Go 150 light transport and trainer and the Go 242 cargo glider.

This rare image shows the Horten IX V2 accelerating for takeoff at the Oranienburg airfield. Photo: courtesy Peter F. Selinger.

Göthert thought it would be better to mount the engines over and under the wing central section on the outside of the wings, what would almost certainly create much more drag. He was also not very confident about the directional stability of the plane, so he installed small vertical fins near the wing tips.

This aircraft never had the chance to prove its superiority (or not) over the Horten brothers design, and the end of the war made it irrelevant.

The Ho IX V3 was almost completed when it was discovered by Allied troops and promptly aroused the interest of its captors. The prototype was sent to the United States and the possibility of finishing it and putting the plane to test was even considered, but the idea was soon abandoned. The prototype was preserved and today is part of the collection of the National Air and Space Museum, where it awaits restoration.

At the end of the conflict, the Horten brothers evaluated the possibilities of continuing their research, preferably in one of the Allied nations.

Following their capture, they were taken to London for interrogation. Yet no work contract was forthcoming and the Hortens found themselves unemployed. Because of the temporary banishment of any aeronautical work in the country, there was no way of continuing the development of new aircraft in Germany.

Reimar decide to leave Germany and try a new start in a most unexpected place: Argentina. He sent his résumé to the *Instituto Aero-Técnico*, in Córdoba, and a short time later arrived in the country to resume his career in the same manner as in the years that preceded the war.

Argentina became a safe haven for former Nazi officials as well as many German aeronautical specialists searching for new opportunities in which to continue their work. One of the first to arrive was Kurt Tank (see "T" tail section) in 1947.

Promptly after his arrival, Reimar enrolled in a sailplane club, the Condor. His design career got a new start with the design and construction of a family of hi-performance sailplanes. Two of them are preserved at the *Museu Aeronáutico*, in Buenos Aires: the IAe 41 *Urubu* and the *Piernifero*.

The IAe 41 was designed in response to a need by many glider clubs for a training sailplane with two seats, side-by-side in this case. Its span was 18 meters and was very easy to pilot. Its major conquest was a historic flight in which Heinz Scheidhauer (the Horten brothers' test pilot since 1938) crossed the Andes from Argentina to Chile, on October 30, 1956.

The 1949 *Piernifero* can be described as the Argentine version

The *Piernifero* without its plywood skin, as it is displayed at the *Museo Aeronáutico* in Buenos Aires (Morón).
Photo: Claudio Farias.

Luftwaffe - confidential - Fundamentals of modern aeronautical design

The sole prototype of the IAe. 38 *Naranjero* (orange carrier). Photo: courtesy Peter F. Selinger.

Artist's impression of the Horten VIII. Illustration: Daniel Uhr.

of the hang-glider. An extremely light and cheap glider (about US $100, at the time), that could be launched by the pilot without the need of a towing plane. This is same way it happens today with the pilots who jump from Pedra Bonita, in Rio the Janeiro.

In a certain way, it was just like going back to German pioneer Otto Lillienthal's time. Lilienthal built an artificial hill around 1894, from which he took off and made several flights.

Reimar Horten also designed a four engine transport plane, the IAe 38 (a further development of the six engine Ho VIII, designed during the war), specifically to transport oranges from the Argentine hinterland to Buenos Aires. For this reason it was called the "*Naranjero.*"

For political and prestige reasons the plane was equipped with low power Argentine engines. This made it suffer from the worst affliction an aircraft can have: impotence. It flew for the first time in 1960 (ten years after the project started) and made just

three more flights before it was abandoned and scrapped.

Mention should also be made about Reimar Horten's project for a flying wing fighter, designed to compete with Kurt Tank's *Pulqui II* (see "T" tail section).

Reimar Horten initiated the design in 1949 and planned to use some components of the *Pulqui II*, (cockpit and landing gear) in order to cut costs and have a prototype completed as soon as possible.

As with other Horten aircraft, the project would follow the usual method of building a glider version to test the aircraft at low speeds.

The Tailless Fighter project, as it became known, employed a double delta configuration, with a 58° sweep in the inboard section of the wing and 40° in the outboard sections. It also featured double vertical rudders positioned over the wings. This layout would provide good flying characteristics at high subsonic speeds, while also permitting low landing speeds.

Reimar Horten planed to use a Rolls Royce *Derwent V* engine which would be fed by an air intake positioned under the fuselage, just as in the Lockheed/General Dynamics F-16.

With this engine, it was calculated that the fighter would have better performance than Kurt Tank's *Pulqui II*.

The project was cancelled in April 1951, just two months after the *Pulqui II* was officially introduced, and never had the chance to show its merits.

The list of Reimar Horten's projects in Argentina also includes the IAe 37, a glider built to evaluate the low speed flight characteristics of aircraft with delta wings. There was also the IAe 48, a reconnaissance delta wing plane, powered by two Rolls Royce *Avons*, capable of reaching Mach 2.2. This project was developed using the Horten brothers' well known method of using free flight models. This time

Artist's impression of the Horten jet fighter. Illustration: Daniel Uhr.

The IAe 37 featured a prone pilot cockpit, but it was later rebuilt to incorporate a conventional cockpit.
Photo: Courtesy Reimar Horten via Peter F. Selinger.

Reimar had the additional benefit of a wind tunnel to help him in the preliminary design stages.

Argentina's grand aeronautical dreams never came to fruition and this project shared the same destiny as Kurt Tank's *Pulqui II* by being cancelled in 1959 on an account of a lack of funds.

TWIN FUSELAGE AIRCRAFT

Germany also produced a few projects that foresaw the use of twin or double fuselages, a configuration whose origins can be traced back to WWI.

Initially the justification for the adoption of such a configuration would be the necessity of providing the necessary distance between propellers in a twin-engine airplane, or between the propellers and a central pod carrying the pilot's cockpit (located between the two fuselages).

Among the first aircraft to use this solution, during the 1914-18 conflict, were the German Fokker K-1 and the Italian Caproni series of bombers. In the period between the two world wars the twin-fuselage concept was successfully used in the Savoia Marchetti S-55 flying boats of which a single example is preserved in São Paulo.

During WWII, practically all nations designed and built twin-

The flying laboratory DFS 332, to be used to test wing sections.
Illustration: Daniel Uhr.

The unusual North American F-82 *Twin Mustang*. Photo: courtesy USAF.

fuselage aircraft, such as the the British *Twin Hotspur* cargo glider, the Dutch Fokker G1 fighter and the Italian SIAI Marchetti SM 91 and SM 92 fighters.

Close to the end of the war, the Americans designed the XP-82 *Twin Mustang*, which better illustrates the twin-fuselage concept. At first sight it resembled two P-51 *Mustang* fuselages joined by a central wing section and horizontal stabilizer. In reality, only a few original P-51 components were used in this conversion and the *Twin Mustang* was not ready in time to be used during the 1939-45 conflict. It did have a decisive participation in the Korean War, especially in its night-fighter version which carried an enormous pod installed under the wing's central section, which housed the radar and machine guns.

In Germany, a series of designs foresaw the twin fuselage configuration, among them the DFS 203 transport glider, created from the union of two DFS 230 fuselages, but it did not reach the production stage.

Another intriguing project was the DFS 332, which was specifically designed as a tool to test real scale wing sections, a sort of flying wind tunnel. The space between the two fuselages would be used for the installation of the wing section to be tested. It would be possible to modify the test wing's angle of incidence in flight in order to evaluate its profile in many flight regimes.

The DFS 332 would be powered by Walter rocket engines that would allow tests in a broad speed envelope. Yet even though the construction of a prototype had been ordered, it was never finished.

The best known German twin fuselage aircraft was the Heinkel He 111Z, a strange creature built by mating two He 111 bombers. These were joined by a central wing section, which also carried an extra engine.

This monster owed its existence to the need for an airplane capable of towing the enormous Messerschmitt Me 321 *Gigant* transport glider, another unusual idea of the Nazi leaders.

The twin fuselage configuration found little favor when it came to jet propelled hi-performance projects, and the most significant example was the Heinkel P-1078B.

This strange design can best be described as a single fuselage aircraft with its nose divided into two pods. The left side pod would house the cockpit while the right side would have plenty of space for the armament and nose landing gear.

Between these two pods it would be an air scoop for the HeS 011 engine. Placing this project in the tailless category of airplanes, this design did not anticipate the installation of a conventional vertical rudder. This is not to mention the asymmetrical positioning of the pilot. All these characteristics make the P-1078B a serious contender for the "Strangest Aircraft Ever Designed in the Third Reich" Award.

After the war the twin fuselage configuration seemed to have been quickly forgotten by aeronautical designers, until the brilliant designer Burt Rutan rescued it from oblivion. In 1986, he used a triple fuselage layout for the *Voyager*, the first airplane to make a round the World flight without stops or in-flight refueling.

ASYMMETRY

From what has been described so far, the only aircraft design dogma still untouched, is the one that says that airplanes are symmetrical objects when seen from above.

The "good manners" aeronautical design manual points out that the right side of an aircraft should be identical to the left, with the probable exception of

Artist's impression of the BV 141. Illustration: Daniel Uhr.

some small details as air inlets, windows or avionic antennas.

Good manners are in short supply during periods of war, so this last bastion of aeronautical normality would soon be knocked down by German designers. As can be expected, the final stroke came from Dr. Richard Vogt's imaginative pencil.

Vogt was the radical apostle of asymmetry as applied to aeronautical design and innumerable aircraft designs foreseeing this arrangement sprouted from his drawing board. What is more surprising in this apparently fantastic story is in the fact that it actually became a reality!

Blohm und Voss designed and built the prototype of an asymmetrical aircraft. When this airplane flew it apparently fulfilled all of its design specifications. It seems that it only failed to reach the operational stage due to Luftwaffe intransigence and conservative thinking, which could not accept the presence of an asymmetrical aircraft in its ranks.

The aircraft in question was the Blohm und Voss BV 141, a reconnaissance airplane which was created to compete with the Focke Wulf Fw 189, designed by Kurt Tank.

Although the BV 141 lost the competition to the Fw 189, Vogt never lost faith in his beliefs and continued to design aircraft with asymmetrical layouts to fulfill the most diverse missions.

With the availability of the new jet engines, Vogt saw a new possibility of installing auxiliary turbojets in a piston engine aircraft in order to increase its performance at special moments of the mission.

From the original configuration of the BV 141, Vogt created a new family of multi-functional projects under the generic category "P-194." The glass covered cockpit of the BV 141 gave way to a metallic nacelle for the pilot (and gunner in the attack versions) and a Jumo 004 turbojet was installed beneath it to provide extra thrust and higher speeds.

TAIL CONFIGURATIONS

Ever since the first heavier than air aircraft flew for the first time more than a hundred years ago, designers and inventors have dedicated a great deal of effort and creativity in the design of the arrangement formed by the rudder and horizontal stabilizers of aircraft.

Many configurations had been tested including simple, double or triple rudders, as well as single or double empennages. The majority of current commercial aircraft use simple rudders, but in military aviation it is possible to find many aircraft that have double rudders, such as in the MiG 29 and Boeing F-15 fighters.

During WWII, other empennage arrangements were developed. Once more this line of investigation sought new alternative configuration, better fit to hi-performance jet aircraft flight regimes. Besides the search for higher speeds, aeronautical engineers were also looking for solutions that would diminish the structural weight of conventional aircraft. This could mean a gain, even a small one, in terms of performance.

BUTTERFLY TAIL

This search for alternatives was the motivation behind a series experiments made by Messerschmitt that would lead to the creation of the so-called "butterfly tail."

Luftwaffe - confidential - Fundamentals of modern aeronautical design

This artist's impression portrays an operational Bf 109G with butterfly tail. Illustration: Daniel Uhr.

In this configuration, the three elements that constituted the tail the vertical fin and rudder as well as the right and left horizontal stabilisers, are replaced by only two "elevons" arranged as in a letter "V."

When the elevons are moved together (movement in the same direction), the aircraft will go up or down. When the elevons are moved asymmetrically (one moves down while the other moves up) it will replace the action of the vertical rudder.

The butterfly tail is also lighter and simpler to manufacture than a conventional empennage, certainly an advantage that cannot be ignored during war, since the rational use of available raw materials is of utmost importance.

Messerschmitt research indicated that the butterfly tail would be especially adequate for high-performance aircraft, thus becoming the ideal choice for the design of jet fighters.

Willy Messerschmitt may have been influenced by a project that was in the final phase of development when France was invaded by German troops, in 1940.

This was the Bugatti 100P record plane, designed by Pierre Louis de Monge. Its prototype was disassembled and hidden on the property of the Bugatti family to prevent its capture by German troops.

It featured an extremely original configuration, which included many innovative solutions. These are seen in the cooling system, the flaps and the "V" tail, all of them duly patented.

The butterfly tail of the Bugatti 100P served to support the air intakes for the radiator, which was located inside the fuselage. In contrast to Messerschmitt designs, the 100P also had a ventral rudder.

So it can be better described as an aircraft whose horizontal stabilizer possessed an accentuated inclination (dihedral), or more appropriately, an aircraft with a "Y" tail and not a pure butterfly tail (without the vertical fin).

In order to evaluate the performance of the butterfly tail in real world conditions, Messerschmitt decided to modify a Bf 109 fighter. He did this by removing its original empennage and substituting it for a set of two surfaces set at an angle of 35^0 iin relation to the longitudinal axis.

This special prototype (V48) was a "G" series aircraft (construction number 14003 and registered VJ+WC), and was tested by the famous company test pilot, Fritz Wendel, in 1943.

Tests demonstrated very small performance gains in relation to the series aircraft and therefore it was decided not to adopt the butterfly tail in the Bf 109.

It is important to point out that this decision was justified more

Aft view of the Bf 109 V48. Photo: courtesy Franz Selinger.

on account of the difficulty in modifying the existing production lines, than for an inherent design flaw of the butterfly tail.

The disappointing performance gains could also be attributed to the fact that the new tail configuration was mated to an existing design which was initially conceived in 1935, and therefore was at the end of its useful life.

The innovative characteristics of the butterfly tail would be used to greater advantage in a project generated from a clean sheet of paper, or applied to high-performance aircraft designs, as was the case in the legion of jet aircraft designs that flowed from the company's drawing boards.

Messerschmitt demonstrated his faith in the butterfly tail concept when proposing its adaptation to some aircraft that were reaching test or manufacture stages. These included the Me 163 Komet, the Me 262, as well as others that were in the initial design stages, such as projects Me P-1101 and Me P-1106.

The concept was also adopted by other German manufacturers, and the best example is the Heinkel 162 *Volksjäger*. Although the series version of this aircraft used a double rudder, later versions were considered with wings having positive or negative sweep. Both versions anticipated the substitution of the conventional empennage with butterfly tails.

Before the end of the war, Heinkel initiated the development of a jet night fighter project known as the P-1079A. The project evolved at such a slow pace that it was still in a very basic stage of development when the war ended.

Curiously, Heinkel designers were told to continue the development of the project in the summer of 1945. Drawings were finished by order and under the supervision of American military forces, in Landsberg, Bavaria.

After the war, the "V" tail concept would also be investigated by the winning nations, especially the USA.

The butterfly tail was not a completely new concept in America; it had been investigated in the 1930's by Bell's future company engineer, Robert Stanley, in an experimental glider he had designed.

Later Bell adapted a P-63 *Kingcobra* (an improved version of the famous P-39 *Airacobra* fighter) to receive this type of empennage, but just as it was with the Me 109 V(48), test results seemed not to justify the introduction of the "V" tail in the aircraft. The idea was rescued in 1948, this time with an eye towards its application in high-performance aircraft. Another *Kingcobra* was modified, but tests once again proved disappointing.

Luftwaffe - confidential - Fundamentals of modern aeronautical design

The photo shows two versions of the exotic Republic XF-91 *Thunderceptor* interceptor fighter. The aircraft on the left displays a search radar inside the radome over the engine intake and on the right can be seen the prototype with the butterfly tail. Photo: courtesy USAF.

Everything seemed to lead to a belief that the butterfly tail would turn itself into one of the countless dead-ends of aviation history. Then everything changed when the configuration was seriously considered for application in the first generation of American experimental supersonic aircraft, the "X" series.

Once again Bell was the driving force behind these ideas, which now included a consideration to install a butterfly tail on the Bell X-1 research aircraft (the first airplane to break the sound barrier in level flight, in October 1947).

Wind tunnel tests with scale models of the X-1 confirmed the viability of the "V" tail, but initially these were not followed by tests with full scale airplanes.

Republic also applied the butterfly tail in one of its more advanced projects, the Republic XF-91 *Thunderceptor*. This interceptor fighter incorporated a bizarre aerodynamic configuration in which the wing chord (distance between leading and trailing edges) increased from the root to the tip, while the same thing was happening with the wing thickness. Moreover, wing incidence angle could also be altered in flight.

The same solution was used in the empennage. The experimental fighter also used a mixed propulsion system, composed of a turbojet engine and a Reaction Motor rocket engine with four combustion chambers.

Later the United States Air Force insisted that the fighter be modified in order to incorporate the "V" tail, believing that this would diminish the aerodynamic drag.

Beechcraft *Bonanza* ad published in Flight magazine, June 1961.

The Fouga *Magister* displaying the colors of the *Esquadrilha da Fumaça* (Smoke Squadron – Brazilian Air Force's demonstration team) preserved at the MUSAL in Rio de Janeiro. Photo: Daniel Uhr.

The XF-91 was duly modified and tests were validated with wind tunnel data. A reduction in drag and an increase in the final speed was clearly observed. Unfortunately the results were not good enough to guarantee a production contract for the XF-91. Its high performance interceptor mission was put in the hands of fighters such as the Convair F-102 *Delta Dagger,* curiously another aircraft to make use of wartime German technology.

The definitive success of the butterfly tail would only come when it was used in an aircraft of much more modest performance, but an airplane that holds an aviation record that still had not been exceeded.

In 1947, Walter Beech presented an airplane that would transform the executive aircraft market. It was the Beechcraft *Bonanza,* or should we say still is since this aircraft remains in production following the incorporation of Beechcraft into the Raytheon group. The most noticeable aspect of this elegant airplane was, without a doubt, its "V" tail. If production numbers and longevity are adequate criteria to evaluate the success of an aircraft, then the *Bonanza* must appear in first place on any list.

The butterfly tail version, known as Model 35, remained in production from 1947 to 1982, which is a very rare thing in aviation history. The "V" tail was later replaced by versions with conventional tails and by 1995, its production had already reached around 17,200 units.

There existed only one other aircraft that has remained in production for the same length of time as the *Bonanza*. It is the Yakovlev Yak-18 Russian trainer, whose production also started in 1947.

This tail configuration would also attract the attention of many designers in Europe, not only on the continent but also in England.

The French were among the first to flirt with the butterfly tail, applying it in a series of experimental aircraft known as C.M.8.R, manufactured by Fouga in 1949.

Artist's impression showing the Supermarine 509.
Illustration: Daniel Uhr.

The C.M.8.R evolved from a glider, in which a small turbojet was mounted over the fuselage just as with the Heinkel He 162.

For the same reasons that had justified the adoption of a double empennage in the German aircraft, the French also opted for a "V" tail in order to prevent the engine exhaust from hitting the empennage.

In 1951, Fouga carried out a very strange experiment. It joined two C.M.8.R aircraft by a central wing section. It also joined the butterfly tail tips, thus creating a most original tail in "W" form.

The C.M.8.R would lead to the Fouga *Magister,* considered the first basic jet trainer. The prototype made its first flight in July 1952 and this aircraft had the distinction of being operated by Brazilian Air Force's Demonstration Team (better known as *Esquadrilha da Fumaça*, or Smoke Squadron).

The Fouga *Magister,* then manufactured by Aerospatiale, wore the colors of the *Esquadrilha da Fumaça* from 1969 up to 1974 with seven aircraft having been acquired. One of them is preserved at the *Campo dos Afonsos* Museum, in Rio de Janeiro.

In England, answering to a Royal Navy request for a high performance ship borne fighter, Supermarine also designed aircraft that used the butterfly tail.

Supermarine Types 508 and 529 differed in small details and both had butterfly tails as a distinctive aspect, but the concept did not find its way into production aircraft. The Supermarine 529 had its "V" tail replaced by a conventional empennage and received swept wings, thus was transformed into the Supermarine *Scimitar* prototype.

HANS MULTHOPP
and the "T" tail.

Messerschmitt was not the only German company to undertake state of the art research in the field of new tail configurations during the war.

Focke-Wulf was one of the last German companies to embrace the concept of jet aircraft using advanced aerodynamic features. Its first jet aircraft designs were marked by a conservative approach which generally foresaw the installation of jet engines into aircraft whose designs were very similar to the configurations found in piston engine aircraft.

One such design considered was the complicated adaptation of the versatile Fw 190 fighter, whose piston engine would give way to a turbojet designed by Focke-Wulf. As seen in the previous chapter, there were also plans to install swept wings on the Fw 190.

This may explain to a large extent the exclusivity of Heinkel, Messerschmitt, Junkers and Arado in the development of jet aircraft for the Luftwaffe.

Luckily for Focke-Wulf, it had a brilliant aerodynamics spe-

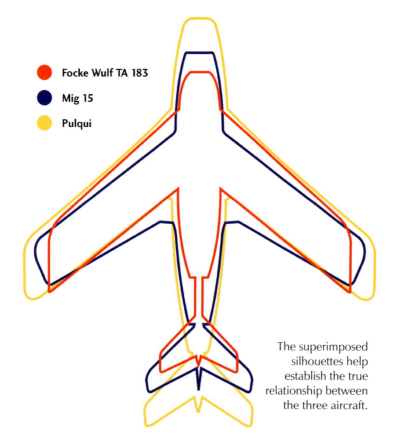

The superimposed silhouettes help establish the true relationship between the three aircraft.

Artist's impression of an operational Ta 183. Illustration: Daniel Uhr.

cialist in its design team, Hans Multhopp.

In 1937, he was studying for his PhD at the University of Göttingen, under the tutelage of Professor Ludwig Prandtl, the father of modern aerodynamics. Prandtl considered Multhopp to be the most talented student of aerodynamics he had ever known.

At this same time Multhopp had obtained a job in the *Aerodynamische Versuchsanstalt* (AVA) in Göttingen, one of the most important aerodynamic research institutions in Germany. At the age of 25, Multhopp was already responsible for one of the institution's wind tunnels, an undeniable evidence of his talent.

In 1938, he was offered a job opportunity at Focke-Wulf, so he quit his theoretical work and did not finish his PhD.

Two years later Multhopp was already Kurt Tank's assistant, who at that time was the company's technical director. In 1943, Multhopp was appointed chief-designer and soon was assigned the task of designing a single-engine jet fighter destined for entry in the RLM competition to develop the so-called Emergency Fighter Program.

This project would come to be known as the Focke-Wulf Ta 183 and would incorporate the brilliant "T" tail solution, which included the installation of the horizontal stabilizer on top of a swept fin, a doubly innovative solution.

While at the AVA, Multhopp perceived that a conventional tail configuration would be far from ideal for the new generation of jets. The positioning of the horizontal stabilizer at the top of the fin increased the momentarm, which would lead to better control at critical moments during take-off and landing.

In aeronautical circles, the "T" tail also became known as the

The IAe. 27 *Pulqui I* preserved at the *Museo Aeronáutico in* Buenos Aires (Morón).
Photo: Claudio Farias.

The photo shows the glider version of the IAe 27 *Pulqui II*. Photo: *Fábrica Militar de Aviones* via Santiago Rivas..

The last surviving example of the *Pulqui II* preserved in Buenos Aires. Photo: Claudio Farias.

"Multhopp Tail," a fit homage to the young designer's talents.

Focke-Wulf received a manufacturing contract for the Ta183 and, apparently initiated construction of a prototype, but the end of the war prevented the plane from being finished. However this would not be the end of Multhopp's fighter.

This tale would have a new beginning in an apparently unexpected place: Argentina.

During WWII and the years that followed, Argentina was governed by General Juan Domingo Perón, a ruler who did not hide his sympathies for the fascist regimes in Europe, especially the government of Adolf Hitler.

As previously seen, after the war, with the connivance of the Argentine government and of many other authorities, Argentina became a safe haven for many Nazi criminals seeking ways to avoid justice at the Nuremberg Court. The most notorious example is Adolf Eichmann, one of the architects of the Holocaust, who was spectacularly kidnapped by the *Mossad* and brought to Israel for trial and execution in 1961.

Furthermore, for many former-members of the Luftwaffe, Argentina offered a new place to work and a chance to reorganize her airforce on the model of the Luftwaffe. Among the aviators who came to the country were General Adolf Galland and Commander Hans Ulrich Rudel, the greatest *Stuka* pilot of the war.

However, it is in the field of technology transfer that can be found the most evident contact between German aeronautical technology and its application by a country not directly involved in the conflict.

While allied countries had to undertake a harsh contest to gain access to German data and scientists, sometimes resorting to conscription (England, USA, and France) or plain simple capture (Soviet Union) of these specialists, Argentina received them freely and spontaneously (in exchange for a calm life free of persecutions, which could not be guaranteed if they had remained in Europe).

Kurt Tank managed to leave Germany in 1946, with the aid of Argentine diplomats and using a false passport.

Focke-Wulf's technical director took with him a microfilm containing the manufacturing drawings of the Ta 183.

Tank also brought with him some members of the original Focke-Wulf team of specialists, including Wilhem Bansemir, Paul Klages, Karl Thalau, and Ludwig Mittelhüber, all of them designers. The group also included Gotthold Mathias, Herbert Wolff, and Otto Pabst.

The group of German citizens that migrated to Argentina also included Reimar Horten and chief test pilot Otto Behrens. Surprisingly Tank's team did not include Multhopp who, as it will be seen later, chose other paths for his career.

For Perón, the presence of the German specialists in Argentina represented a unique opportunity, which could not be wasted.

They could contribute to a decisive leap forward in Argentine

aeronautical technology, placing the country in a position of prominence (also in military terms) in the South American context.

Argentina had already taken a significant step when contracting the services of another expatriated European, French engineer Emille Dewoitine. He left France accused of collaboration during the war and after arriving in Argentina, initiated the development of the IAe 27 *Pulqui* (Arrow, in native dialect). This important airplane was the first jet aircraft designed and manufactured in South America, a pioneering feat, since it made its first flight in August 1947.

The *Pulqui I*, was an extremely important aircraft for the Argentine Air Force's technological renewal program, but its design was completely conventional.

Argentina could dream of much higher and more ambitious flights after Tank's arrival.

Armed with a complete set of plans for the Ta 183, Tank decided to revive the project and at the same time bring it up

The *Pulqui II*, carefully restored and preserved at the *Museo Aeronáutico* in Buenos Aires (Morón). Its Rolls Royce jet engine can be seen in the foreground. Photo: Claudio Farias

to date, especially in what concerned the propulsion unit. In its original incarnation, the Ta 183 anticipated the installation of axial turbojets, such as the Jumo 004 or the Heinkel-Hirth HeS 011, but to manufacture or even obtain these engines in Argentina was totally out of the question. This led to the decision to choose a Rolls-Royce *Nene* centrifugal flow engine as substitute.

This would generate profound alterations in the design. Yet the final blow came when Tank decided to change the position of the wings in relation to the fuselage, mounting them as shoulder wings instead of the mid-fuselage position of the original design.

These modifications transformed the flight characteristics of the aircraft, making it difficult and even risky to fly. Argentine pilot Edmundo Osvald Weiss would bear witness to this while carrying out the inaugural flight of the airplane, now baptized *Pulqui II*, in June 1950.

Project errors became evident prompting the prototypes to be constantly modified. Accidents destroyed two prototypes and caused the death of two pilots, among them veteran Otto Behrens. It was claimed that the accidents were not caused by inherent design flaws, but in the end, this would lead to the cancellation of the project after the construction of only five prototypes (and one static test specimen).

The project also became excessively costly for a developing country such as Argentina. At that time it was possible to acquire a used F-86 *Sabre* for the same price of a new *Nene* turbine.

FURTHER PROJECTS IN ARGENTINA

Although the primary mission of Tank's group in Argentina was the development of a high-performance jet fighter, the *Pulqui II*, at the same time he also initiated the design of an advanced jet airliner. The IAe–36 *Condor II,* so named as to allude to Tank's first successful airliner, the pre-war Fw 200 *Condor.*

The first conceptual drawings of the *Condor II* were done in 1948, although some sources claim the design was born in Europe just after the war, and called Ta 500. The design evolved into a very modern aircraft, configured to carry 31 to 36 passengers and with an estimated range of 5,000 km, flying at 950 km/h.

In its initial configuration, the *Condor II* displayed wings swept at 35⁰ and just as its smaller brother, a swept (45⁰) "T" tail. The aircraft was powered by five Rolls Royce *Nene II* turbojets, clustered inside the

Artist's impression showing the *Condor II* in the colors of Aerolineas Argentinas. Illustration: Daniel Uhr.

tail, fed by a circular intake, and exhausting through a common central pipe.

The *Condor II* showed a marked resemblance to the SE.210 Sud *Caravelle*, whose design was initiated at approximately the same time.

The *Condor II* reached the wind tunnel test phase and it seems that a prototype was ordered in 1953, but construction was halted soon after.

A curious note was published in the Brazilian aviation magazine *Aviação*, in June 1953, regarding this project and Kurt Tank's high hopes for the *Condor II*: "According to Mr. Tank, who lives in Argentina since the end of the conflict, on the day his jet airliner is put in series production, that country will assume a leading position among aircraft building nations."

Although this book mainly deals with jet projects developed in Germany, it should be mentioned that Kurt Tank's group in Argentina also designed more mundane, but nevertheless clever, aircraft for the Argentine Air Force.

A Focke-Wulf Fw 58 *Weihe* of *Sindicato Condor*. This aircraft, named *Cacuri* was used by the Brazilian airline from 1940 onwards. The main difference from the military version is the solid nose instead of transparencies present in the bomber-trainer version used at the same time by Brazilian Naval Aviation. Photo: Claudio Farias' collection.

The IA 35 *Huanquero* preserved at the *Museo Nacional de Aeronáutica* in Buenos Aires. Photo: Claudio Farias.

One of these projects was the IA 35 *Huanquero*, designed by Paul Klages and Kurt Tank, circa 1950, and which flew for the first time in 1953.

This project spawned a whole family of aircraft designed to fulfill the following missions:
Version IA would be a trainer aircraft for pilots and navigators, carrying an instructor and four pupils.
Version IB would be an armed trainer, featuring two 12.7 mm Browning machine guns in the nose, two more mounted in an hydraulically operated turret, a Mk 14 bomb sight, as well as racks for two 100kg or four 50kg bombs. Rockets could also be carried under the wings.
Version II would be used as a light transport, carrying seven passengers.
Version III would be an air ambulance, with four stretchers and one medic.
Version IV would be used for photographic survey, with a Fairchild 225 camera and one operator.

It is impossible to read this description and not relate this project to a previous, equally versatile design by Kurt Tank, the Fw 58 *Weihe*, which was designed in 1935 to fulfill the same kind of tasks reserved for the *Huanquero*.

In fact, as the *Weihe* was a familiar sight in South America, the *Weihe II* designation would not sound out of place regarding this pedestrian, but certainly very useful aircraft. Three Fw 58 were exported to Argentina in 1938 and it was also built under license in Brazil. The last restored example now resides in the *Museu Aeroespacial*, in Rio de Janeiro.

The *Huanquero* was one of the few German projects that survived the political turmoil in Argentina. In fact it prospered under these unfavorable conditions. Further developed by Argentine engineers, it evolved into a series of successful aircraft culminating in the IA 50 G-II, which was powered by turbo-propeller engines and featured a swept fin.

This airplane made its first flight in 1963 and 34 aircraft were built, used mainly by the Argentine Air Force but also by private users. Many of them were still in use 25 years later, making the *Huanquero* family one of the most successful examples of German technological transfer in Argentina.

Argentina's political instability was the main factor conspiring against the work of the German specialists that came to that country.

In September 1955 Perón was overthrown and the new government was not as receptive to the German engineers and soon

The final development in the *Huanquero* family, the turbo-prop IA 50 G-II, seen at the *Museo Nacional de Aeronáutica* in Buenos Aires
Photo: Claudio Farias.

many left Argentina, including Kurt Tank, who went to India in 1956.

Ta 183 DESCENDANTS

The Argentines were not the only ones to be impressed by the potential of the Ta 183. After capturing Berlin, the Soviets took possession of a vast amount of aeronautical documents that had been diligently filed in the RLM.

These documents included complete plans of the Ta 183. These were carefully studied by Soviet engineers who concluded that a series of modifications could be incorporated into the original project, so as to improve its flight characteristics at low speeds.

In order to confirm the captured German data, Soviet specialists carried out a series of wind tunnel tests at TsAGI, (Central Institute of Aerohydrodynamics) which had been recently modernized with captured German equipment and data.

Pulqui II, Walkaround. *Museo Nacional de Aeronáutica* in Buenos Aires. Photos: Claudio Farias.

The Soviet design bureau of *Artem Mikoyan* and *Mikhail Gurevich* (better known in the west by the MiG acronym), decided to add wing fences (see CHAPTER ONE), which are vertical metallic plates positioned at intervals along the wing chord in order to prevent span-wise airflow. The plane also featured wings that drooped slightly downward (anhedral).

Wind tunnel tests also pointed out the necessity of repositioning the horizontal stabilizer, until then mounted at the top of the fin. Once again motivated by the quest for better performance at low speeds, the stabilizer was placed in an intermediate position, approximately half way between the base and the top of the rudder.

The fruit of this German/Soviet marriage flew for the first time in December of 1947, and became known as the MiG 15. Its performance in the Korean War, and the sheer amount of concern it caused among western military experts, is the undisputed testimony of its success.

It would be untrue and naive to claim that the MiG 15 was a

Luftwaffe - confidential - Fundamentals of modern aeronautical design

The SAAB J29 *Tunnan*. Photo: courtesy David Ilott.

During the Cold War the Soviet counter intelligence services did their job very well, as can be confirmed by this illustration published by Aviation Age magazine in 1951. It portrays the mythical "MiG 19" inspired by the Ta 183.

copy of the Ta 183, but at the same time, it can not be denied the clear influence of the German design in the Soviet fighter. This is especially evident when one compares the layout of previous MiG jet fighters (the MiG 9 of 1946) with the advanced design of the MiG 15.

For some time after the end of WWII, rumors had it that the Russians had built prototypes of the Ta 183. Some illustrations were even published in the specialized press showing a fighter with the same characteristics, marked with the Soviet Air Force red star.

These fancy illustrations served as the starting point for the now defunct plastic kit manufacturer Aurora, who released its spurious version of the "MiG 19" in 1:48 scale. This model, injected in a green-metallic color, added a nose cone over the original air intake of the Ta 183, suggesting the installation of interceptor radar as on the North American F-86D *Sabre Dog*.

The Ta 183's DNA circulated freely in many parts of Europe, and Sweden demonstrated a special interest in its innovative characteristics.

SAAB, who manufactured some German aircraft (Junkers Ju 86 and Focke-Wulf Fw 44) under license before the war, and had established commercial relations with German aeronautical related industries (Daimler Benz), came to know about the existence of a series of Luftwaffe research reports in Switzerland about swept wings.

Apparently, these reports failed to arouse the interest of Allied authorities, presumably because of the great number of pages to be translated.

For the Swedish, the German language was not a barrier and the content of the reports proved to be a gold mine of information on advanced aerodynamic configurations.

The data was promptly applied in the project of a fighter, initiated in 1945, that would come to be known as the SAAB J29. The Swedish team of designers included engineer Behrbohm, a former structural analysis specialist at Messerschmitt.

SAAB engineers opted for a cautious approach and decided for a moderate 25⁰ wing sweep for the airplane, which would be powered by a British de Havilland *Ghost* turbojet.

Seen from above or from the side, the SAAB J29 shows "slightly obese" contours and silhouette, and this prompted Lars Brising, head of the design team, to call it *Tunnan* (Barrel). Yet this pregnant version of the Ta 183 hid outstanding flight characteristics behind its rather portly appearance, as was confirmed after its first flight in 1948, as well as by later tests when the aircraft broke speed records, until then in the possession of the F-86 *Sabre*.

To be absolutely precise, the *Tunnan* owes more to German research in terms of its wing

The transonic research plane designed by Multhopp in England.
Illustration: Daniel Uhr.

The Martin XB-51 featured many characteristics previously seen on WWII German designs: "T" tail, bicycle landing gear, variable incidence wing and last, but not least, turbojets mounted at the fuselage sides, just like in the Junkers Ju 287. Photo: courtesy USAF.

design, than to the general arrangement of the Ta 183. Although *Tunnan's* layout is quite similar to the one adopted by the Focke-Wulf fighter, its empennage is totally different.

Given the presence of a former Messerschmitt engineer in the design team, it can also be argued that the J29 could also have been influenced by the Me P-1101 project.

While the spirit of the Ta 183 was reincarnated in aircraft from all corners of the world, Hans Multhopp found new job opportunities in England.

He was invited to work at the Royal Aircraft Establishment

This is one of the original SV-5D/X-23A Lifting Body test aircraft, preserved at the USAF Museum. Photo: courtesy USAF.

(RAE), in Farnborough, Great Britain's most important aeronautical research center.

At that time, the international aeronautical community, especially in the USA and England, was taken by a frenzy caused by the recent discovery of the results of German research in the field of transonic speed flight.

As previously seen, this line of research spurred the development of the de Havilland DH 108 and the Northrop X-4, both tailless aircraft, a configuration that could result in reduced stability at some flight regimes, as was demonstrated by the DH 108.

The incredible Martin *Seamaster*, an idea too exciting to be forgotten. Illustration by Daniel Uhr over Revell model assembled by Átila Coutinho (in memorian).

For many experts the most efficient way to reach transonic and supersonic speeds would not reside in the elimination of the tail but instead in the use of wings of extreme sweep and small thickness, associated with an equally efficient tail configuration for high speed flight: the "T" tail.

So Multhopp initiated the design of an aircraft with a radical configuration: a tube-like fuselage inside of which the cockpit (from where the pilot would control the aircraft in a prone position) was situated just behind the air intake to the engine. The wings were swept at 55^0, a skid landing gear eliminated the wheels (*a la* Me 163 *Komet*), and finally as could be expected from Multhopp, a "T" tail.

This hot-rod would be powered by a Rolls-Royce *Avon* turbojet and specifications indicated a maximum speed of about 1,300 km/h, an impressive performance for the day.

Unfortunately the project was not continued, being cancelled around 1947.

Multhopp packed and went to the United States, initiating a long and productive period at the Glenn L. Martin Company.

During Multhopp's stay at Martin, the company produced a series of aircraft that used the

An Ilyushin IL-62 (and its prominent "T" tail) from *Cubana*, at Montevideo airport.
Photo: courtesy Otávio Lamas de Farias.

"T" tail, the most notable being the XB-51 attack/bomber and the fascinating jet flying boat, the XP6M *Seamaster.*

Later Multhopp involved himself in the design of the SV-5D/X-23A Lifting Body aircraft, a direct forerunner of NASA's *Space Shuttle* program.

He died in Cincinatti, in 1972.

For aviation enthusiasts there is only one way to solve the controversy around the real potential of the Ta 183, which is considered by some authors to be the most important and influential of all German designs developed during the war: build a prototype of the Ta 183 using a jet engine of equivalent power to the original HeS 011.

For those who think this idea is a little bit too fantastic, it must be remembered that an entirely new series of Messerschmitt's Me 262 and Focke-Wulf's Fw 190, are already flying. These were manufactured by a process known as reverse engineering (in the case of the new Me 262s) or from original manufacturer's drawings.

The "T" tail seems to have found its definitive habitat in commercial aviation, as attested by the great number of airliners that have adopted this solution. This includes such classics as the de Havilland *Trident* (1961), the Vickers VC10 (1962), the Ilyushin IL 62 (1963), the Boeing 727 (1963), the Tupolev Tu 134 (1963), the BAC *One-Eleven* (1963), the Douglas DC-9 (1964), and finally the Fokker F28 (1967).

The "T" tail also seems to have perfectly adapted itself to tropical climes, as evidenced by a series of aircraft manufactured in Brazil by EMBRAER: the EMB-120 *Brasilia* (1983), the EMB-121A *Xingu* (1976), the prototype CBA-123 *Vector* (developed in partnership with Argentina - 1990), and the EMB 145 which was the first jet airliner designed and manufactured in Brazil (1995).

The configuration also became popular among military transport aircraft, as it is the case of Lockheed's C-141 *Starlifter* (1965) and C-5 *Galaxy* (1968), of the Ilyushin Il-76 (1971), and of the Boeing C-17 *Globemaster III* which is currently in use by the United States Air Force.

CANARDS

There exists yet another tail configuration that deserves to be mentioned in this chapter. It is the most unusual of them all as it removes the horizontal stabilizers from the rear extremity of the aircraft and moves them to the nose, usually ahead of the cockpit.

In a certain way this should be the most usual tail configuration since, historically, it preceded all the others.

The two aircraft that managed to make the first controlled flights in the U.S.A. and in Europe, the Wright *Flyer,* designed

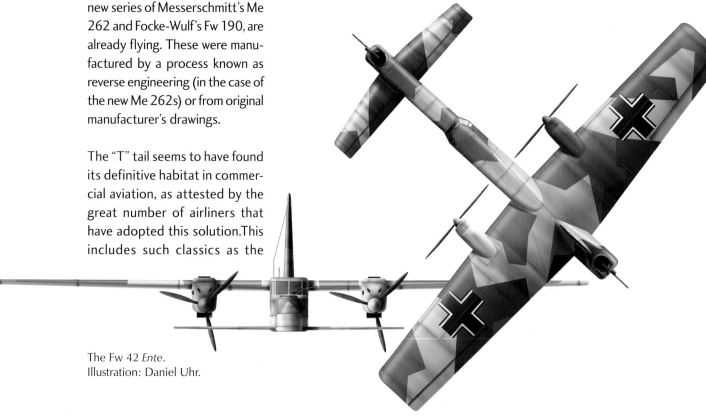

The Fw 42 *Ente.*
Illustration: Daniel Uhr.

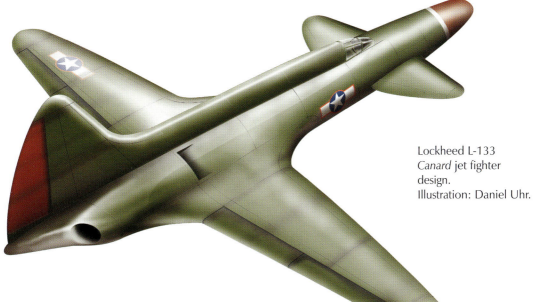

Lockheed L-133 *Canard* jet fighter design.
Illustration: Daniel Uhr.

by the Wright brothers, and Santos Dumont's *14bis*, used this configuration.

In the first decade of aviation history, this arrangement was very popular among pioneering designers. Since then, however, it has been gradually abandoned and replaced by the configuration that is considered standard today: the positioning of the tail at the aft extremity of the fuselage.

The *canard* term, "duck" in French, seems to have been adopted alluding to the look of this bird in flight, where the wings seem to be located in the posterior extremity of its body.

Although the arrangement has lost its popularity since the first successful flights in the beginning of the 20th century, many designers insisted on this solution, primarily when the engine was placed in the tail and pushed the airplane. In the traditional arrangement, the engine is placed in the nose and pulls the aircraft.

By the time of WWI, the *canard* arrangement had already been abandoned as a viable proposal for military aircraft. Only in the 1920's and 1930's would it be seriously considered again.

Here again this pioneering vision cannot be exclusively credited to German designers, and even less so to the high-tech designs from 1944/45. All nations flirted with canards, including Germany.

At the time Germany initiated its rearmament in a clandestine way, the Focke-Wulf company received a contract for the development of a twin-engine bomber, the Fw 42.

Beginning in 1932, wind tunnel tests were initiated and later a fuselage mockup was built, just before the project was cancelled in 1934.

The *canard* arrangement was also proposed by Clarence "Kelly" Johnson, Lockheed's chief engineer and father of such outstanding aircraft as the P-38 *Lightning*, the *Constellation* airliner, the F-80 *Shooting Star* and the U-2 spy plane.

In 1938 he presented the design (US Patent 2,271,226) of the Lockheed Model 27 twin-engine

Artist's impression of the Kyushu *Shinden*. Illustration: Daniel Uhr.

The Messerschmitt P-1110 in its initial *canard* configuration.
Illustration Daniel Uhr.

canard airliner. The aircraft would be powered by two 2,000 hp radial engines and would carry up to 35 passengers by day or with the provision for sleeping berths, 16 by night. Maximum speed was calculated at 459 km/h.

Although it failed to leave the drawing board, it may have been a starting point in the design of subsequent Lockheed *canards*, as will be seen later.

During WW II, the concept of *canard* aircraft was explored by all combatant nations, especially in the design of fighters.

In contrast to the great majority of German projects that remained on the drawing board or in the wind tunnel model stage, some foreign *canard* aircraft had been transformed into hardware. At least one of them was destined for series production, should the war have continued beyond 1945.

In Italy, the Ambrosini company created the SS-4, an all-metal fighter prototype which made its first flight in 1939. Equipped with an engine of only 960 hp, the SS-4 could reach 540 km/h. It made some test flights until being destroyed in an accident, in 1940.

In the USA the formula was adopted by Curtiss–Wright for the XP-55 *Ascender*, a fighter of radical pusher design. Three prototypes were built and two were lost in accidents, causing the American Army Air Force to lose interest (at least momentarily) in the concept.

The triple sonic North American XB-70 bomber. Photo: courtesy NASA.

Paradoxically, in view of the sad history of the XP-55, WWII's most advanced *canard* aircraft design had its origin in the drawing board of the American engineer Natham Price.

In 1940 he began under the auspices of Lockheed, the design of a turbojet engine, the axial flow L-1000. At the same time, he also designed an aircraft of revolutionary design (a *canard*) to use it.

The L-133 project, as it came to be known at Lockheed, displayed forms worthy of appearing on the cover of the famous sci-fi magazine *Amazing Stories*. The aircraft would have been capable of reaching about 965 km/h and an altitude of 16,500 meters, thanks to the thrust of two advanced L-1000 engines.

But the USAAF's conservative thinking did not allow the L-133 to be built and Lockheed decided to adopt a more cautious approach for the design of its first jet, the XP-80 *Shooting Star*.

WWII's most impressive *canard* was the Kyushu J7W-1 *Shinden* (Magnificent Lightning), designed by Captain Masaoki Tsuruno, of the Japanese Imperial Navy.

This project foresaw two development stages: initially the airplane would be powered by a Mitsubishi radial engine mounted behind the pilot's cockpit, and turning a six blade pusher propeller at the end of the fuselage, through a transmission shaft. Later the aircraft would be fitted with a turbojet in order to take full advantage of the *canard* configuration.

Before beginning the construction of a prototype, Captain Tsuruno built a glider that reproduced the configuration of the *Shinden* in smaller scale. This allowed an evaluation the aircraft's flight characteristics at low speeds and proved the viability of the project.

What makes the *Shinden* a unique case among *canards* is the fact that the Japanese Armed Forces ordered it into production even before the first flight of the prototype, which occurred in August 1945, just a few days before Japan's final surrender.

In Germany, the *canard* configuration did not find much favor among the many aeronautical companies. Few were the projects that used this solution and even fewer foresaw the use of jet propulsion. One of them that did was a mysterious Dornier fighter project powered by three HeS 011 engines.

Messerschmitt considered a *canard* version for its P-1110 project, as evidenced by the name *Ente*, hand written in some original drawings.

This fighter would use a HeS 011 turbojet and would be armed with four 30 mm cannons. Yet after wind tunnel tests it was decided to go for a traditional configuration (by German standards) for the P-1110, using swept wings and a butterfly tail.

Other German companies used the *canard* configuration in their designs during the war, only to be employed in piston engine aircraft, or on cargo gliders such as Gotha's P-50 project which featured twin vertical rudders on the wing tips.

In 1941 Henschel offered the RLM a *canard* fighter design, the P-75. It foresaw the installation of heavy armament in the nose and would be driven by a set of counter-rotating pusher propellers behind the fuselage.

The Luftwaffe recommended the cancellation of the project arguing that "pilots would never get used to having the propeller behind their backs and the tail in the front of the airplane."

Only after WWII did *canards* return to the aeronautical designers' menu, especially those involved with military projects.

Canards were used in the *Viggen* family of Swedish fighters, manufactured by SAAB, and were also present in the design of the trisonic XB-70 *Valkyrie* bomber, built by North American.

The Soviets used retractable *canards* in their supersonic airliner, the Tupolev Tu 144. More recently, *canards* returned in strength in the new generation of Sukhoi fighters (Su 33, Su 34 and Su 35).

Tailless aircraft

Naranjero

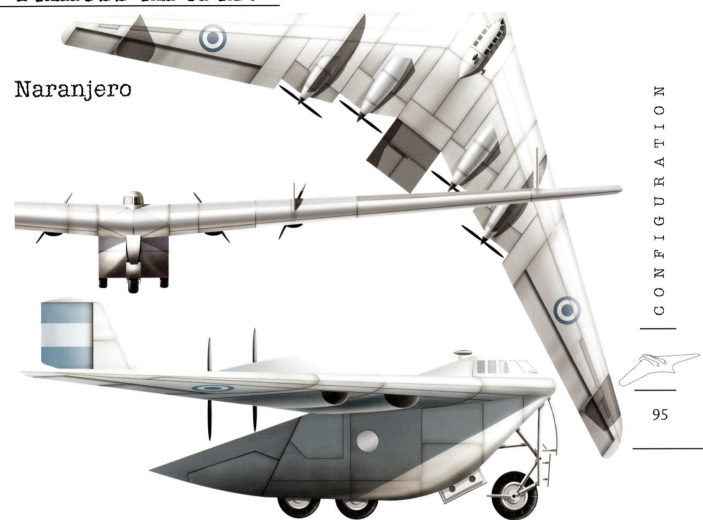

CONFIGURATION

The IA-38 was entirely built of metal and the original design would employ hydraulically actuated drag rudders of the same kind used in the Horten 229. Unfortunately difficulties with the hydraulic system forced Reimar Horten to add conventional rudders mounted on top of the wing. The rear part of the fuselage opened vertically in two parts and could be partially opened in flight in order to deliver cargo using parachutes. The cargo hold had a total volume of 30 cubic meters and the plane could transport up to 8,000 kg.

Data File

Mission:	Transport/cargo airplane
Final development status:	prototype
Engine:	four *El Gaucho* 450HP piston engines
Maximum speed:	252 km/h
Range:	1,240 km
Weight:	17,500 kg (loaded)
Span:	32 meters
Length:	13.40 meters
Armament:	--------------

BV P-208.03

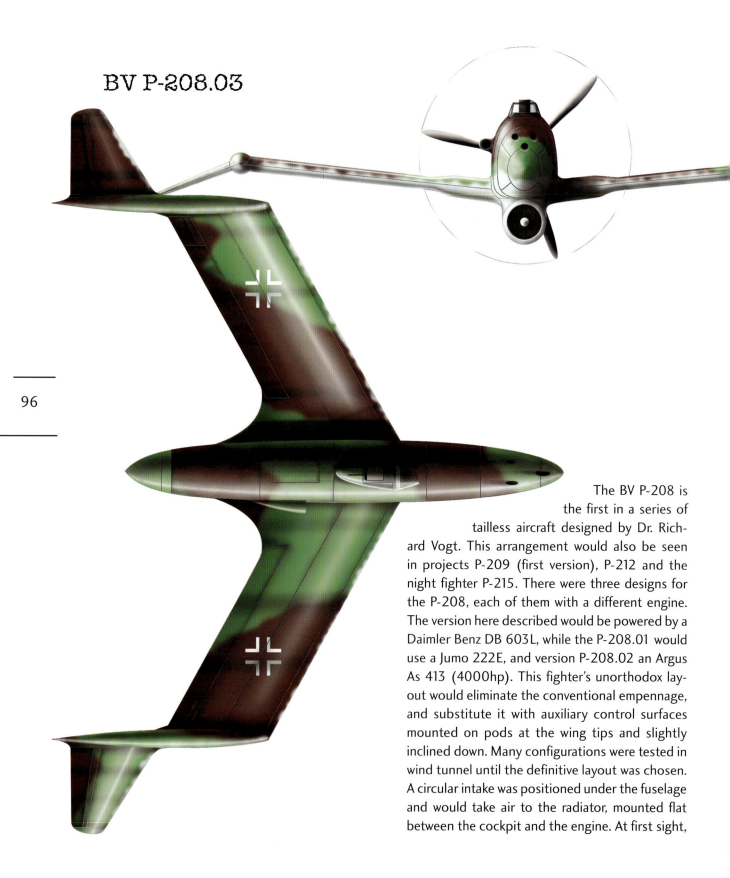

The BV P-208 is the first in a series of tailless aircraft designed by Dr. Richard Vogt. This arrangement would also be seen in projects P-209 (first version), P-212 and the night fighter P-215. There were three designs for the P-208, each of them with a different engine. The version here described would be powered by a Daimler Benz DB 603L, while the P-208.01 would use a Jumo 222E, and version P-208.02 an Argus As 413 (4000hp). This fighter's unorthodox layout would eliminate the conventional empennage, and substitute it with auxiliary control surfaces mounted on pods at the wing tips and slightly inclined down. Many configurations were tested in wind tunnel until the definitive layout was chosen. A circular intake was positioned under the fuselage and would take air to the radiator, mounted flat between the cockpit and the engine. At first sight,

this solution could make one believe that the airplane was jet powered. Post-war British intelligence reports mention an interesting characteristic of the engine's exhaust system. Engine exhaust would escape next to the wing's trailing edge fillet, close to the fuselage and it was thought that this might improve the propeller's efficiency. In order to save strategic raw materials, steel would be employed in 60% of the aircraft's structure. The fuselage would be built in aluminum but the pressurized cockpit would be built using welded steel. The positioning of the propeller behind the fuselage would make an ejection seat mandatory, and also some system to blast off the propeller, in order to provide the pilot with a safe way to leave the aircraft in case of an emergency.

Data File

Mission:	single place day fighter
Final development status:	wind tunnel models
Engine:	one Daimler Benz DB 603 L engine
Maximum speed:	790 km/h at 9000 meters
Range:	1,230 km
Weight:	5,000 kg (loaded)
Span:	9.50 meters
Length:	9.20 meters
Armament:	three Mk 108 cannons

Luftwaffe - confidential - Fundamentals of modern aeronautical design

BV P-212

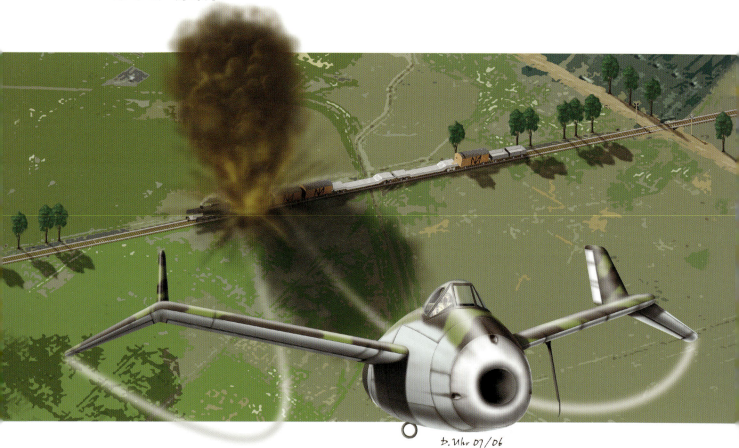

Project P-212 followed the line that began with the piston engine P-208. The design started in 1944 and evolved into different configurations (P212.01 and P-212.02) until reaching the definitive design called P-212.03. It displayed wings with less sweep (40°) compared with its predecessors, and vertical rudders mounted over the wings with downward inclined wing tips, a formula already explored in the P-208. The project looked promising and prototype construction was considered in February 1945.

Data File

Mission:	single place day fighter
Final development status:	wind tunnel models
Engine:	one HeS 011 turbojet engine
Maximum speed:	965 km/h
Range:	-----------
Weight:	4,180 kg (estimated)
Span:	9.50 meters
Length:	7.40 meters
Armament:	From three to seven Mk 108 cannons

Ju EF 128

This project was developed in two versions: a single-place day fighter and a two-place night fighter. It was originally conceived in response to the Emergency Fighter Competition. It was a tailless aircraft powered by an HeS 011 turbojet. It had wooden shoulder wings swept at 45°. The engine intakes would be mounted at the fuselage sides, under the wings. Its configuration is similar, in general lines, to the one adopted by the Chance Vought F7U *Cutlass* carrier borne fighter. The pressurized cockpit included an ejection seat. Models tested in the wind tunnel showed promising results and it was considered for the construction of a fuselage mock-up, including the engine, which was to be mounted piggyback over a Junkers Ju 88 in *Mistel* fashion. Although promising, the project was rejected in favor of the Focke-Wulf Ta 183.

Data File

Mission:	single place day fighter/two place night fighter
Final development status:	wind tunnel models
Engine:	one HeS 011 turbojet engine
Maximum speed:	880 km/h at sea level
	950 km/h at 6.000 meters
Range:	1,800 km
Weight:	4,900 kg (loaded)
Span:	8.90 meters
Length:	7.00 meters
Armament:	two Mk 108 cannons with the option of two more

Luftwaffe - confidential - Fundamentals of modern aeronautical design

Me P-1111

The Messerschmitt P-1111, designed in February 1945, was the penultimate project developed by the company before the end of the war. It tried to solve many of the problems found on projects P-1106 and P-1110. The air intakes, positioned close to the wing roots, would provide a uniform intake airflow and result in less loss of thrust.

This project was abandoned and substituted by the Me P-1112, Messerschmitt's last aircraft designed during the war.

Data File

Mission:	single place day fighter
Final development status:	wind tunnel models
Engine:	one HeS 011 turbojet engine
Maximum speed:	995 km/h
Range:	1,500 km
Weight:	4,281 kg (loaded)
Span:	9.16 meters
Length:	8.92 meters
Armament:	four Mk 108 cannons

Canards

Hs P-75

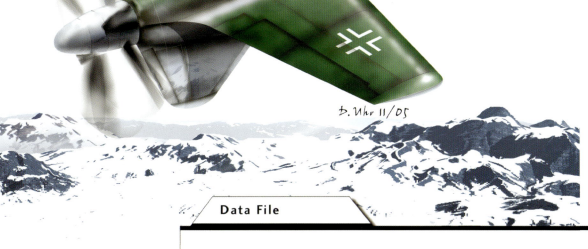

Data File

Mission:	single place day fighter
Final development status:	wind tunnel models
Engine:	one Daimler Benz DB 613 engine (two coupled DB 603s)
Maximum speed:	790 km/h
Range:	150 km
Weight:	7,200 kg
Span:	11.30 meters
Length:	12.20 meters
Armament:	four Mk 108 cannons

Henschel's P-75 project is one of the few applications mating the canard configuration to a piston engine fighter in Germany during WWII. This layout would permit the installation of heavy concentrated armament in the aircraft's nose, but on the other side it would imply the adoption of the same kind of systems employed by the Dornier 335: an ejection seat plus a mechanism that would detach the ventral rudder in event of a wheels up landing. The fuselage cross section design already showed some blending with the wings which made the installation of the huge coupled engine (an arrangement similar to those seen in the Heinkel He 177 bomber) easier. Wind tunnel tests showed promising results but the project was abandoned.

Flying wings

Horten IX/ Horten Ho 229

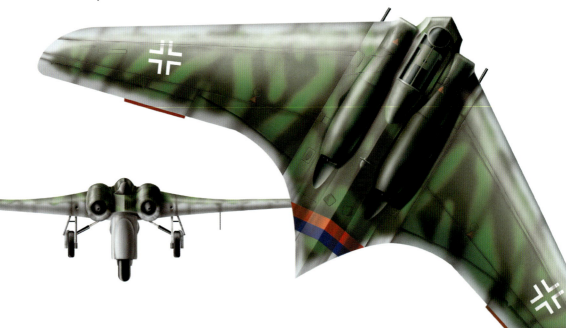

The Horten Ho 229 was the only flying wing fighter to be ordered into production by any nation during WWII and it can be considered the apex in the Horten brothers' career. As in their previous aircraft, the Ho 229 also used plywood as its main construction material. The wing center section was built from a welded steel tube structure and then covered in plywood. Three prototypes were built: initially a glider version (Ho IX V1) to evaluate the aerodynamic characteristics of the project. This was followed by the Ho IX V2, the first prototype with Jumo engines, and which made some flights early in 1945. This airplane crashed and was completely destroyed but, when the war ended, the Ho IX V3 was almost finished and after its capture by allied troops, it was taken to the USA, and is now part of the collection of the National Air and Space Museum, in Washington. The series manufacture of the Horten 229 (official designation of the aircraft) was transferred to Gothaer Waggonfabrik and besides the Ho IX V3, other prototypes, including two seat trainers and night fighters were also under construction.

Data File

Mission:	single place day fighter / two place night fighter
Final development status:	prototype
Engine:	two Jumo 004 turbojets
Maximum speed:	977 km/h
Range:	------------
Weight:	7,726 kg
Span:	16.80 meters
Length:	7.46 meters
Armament:	four Mk 108 cannons or two Mk 103 cannons

Gotha P-60

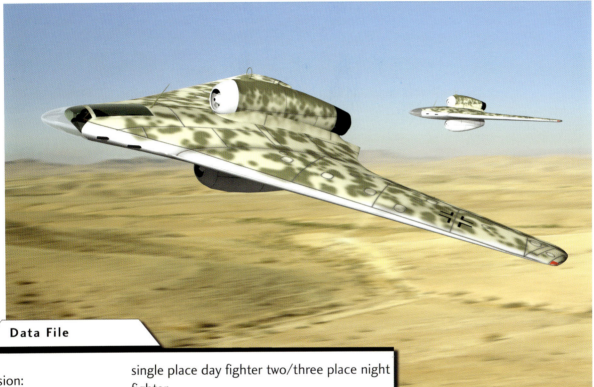

Data File

Mission:	single place day fighter two/three place night fighter
Final development status:	wind tunnel models
Engine:	two HeS 011 turbojets
Maximum speed:	930 km/h
Range:	2,200 km (estimated)
Weight:	10,500 kg (estimated)
Span:	13.50 meters
Length:	10.90 meters
Armament:	four Mk 108 cannons on the wings and two mounted obliquely in the fuselage

This project formed a family of three distinct night fighter aircraft. The first version, known as P-60A was the one that more closely adhered to the concept of a pure flying wing. Pilot and radar operator would fly the aircraft from a prone position, just as in many designs from the Horten brothers. The plane did not feature a vertical tail or fuselage and the engines were located over and under the wing center section. The under wing engine would be prone to damage caused by the ingestion of debris left on the landing strip. Furthermore, this configuration would make a wheels up landing very dangerous. On later versions, the P-60B and C, the design resembled more a tailless aircraft, with a noticeable fuselage and cockpit and vertical rudders installed near the wing tips. The P-60C version would have two extremely inclined seats, mounted inside the wing roots and covered by transparent panels.

Twin fuselage aircraft

Heinkel He P-1078 B

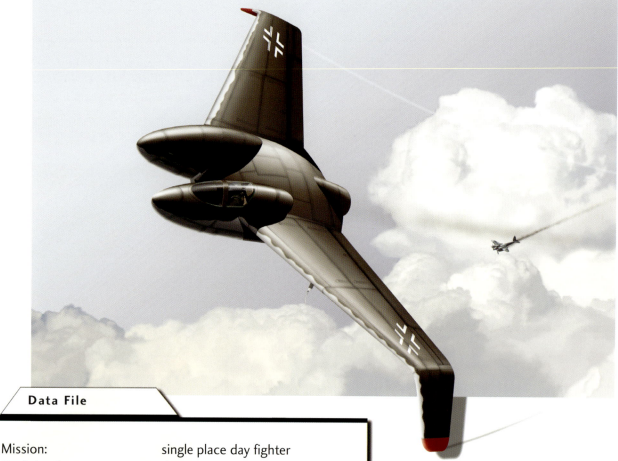

Data File

Mission:	single place day fighter
Final development status:	design
Engine:	one HeS 011 turbojet engine
Maximum speed:	909 km/h
Range:	1,545 km
Weight:	4,037 kg (estimated)
Span:	9.00 meters
Length:	6.00 meters
Armament:	two Mk 108 cannons

Heinkel's project P-1078 was developed in three different versions. Version *A* displayed a conventional configuration while versions *B* and *C* were tailless aircraft. Version *B* was definitely the most original in terms of configuration as demonstrated by its bifurcated fuselage. It was not a twin fuselage in the pure definition of the term as in the Heinkel He 111 Z or even similar to the configuration of the Lockheed P-38 *Lightning*. The frontal part of the fuselage was divided in two pods, with the engine air intake located between them. The left hand pod housed the cockpit while the armament and nose landing gear were located on the right side. It was a very compact fighter with wings swept at 40°.

Assymetric aircraft

BV P 194

This project was initiated in 1944 and foresaw its employment in many different missions: dive bomber (Stuka), reconnaissance, ground attack among others. Since the beginning, this aircraft family predicted the installation of a mixed propulsion system, employing a piston engine and an auxiliary turbojet. The P-194 evolved from similar earlier project, the BV 237, an attack/dive bomber aircraft derived from the BV 141 reconnaissance airplane. The main difference between the P-194 and the BV 237 was the fact that the earlier design, in its attack version, did not incorporate an auxiliary turbojet engine. Another marked difference between the two aircraft was the presence of an asymetric horizontal stabilizer in the BV 237, while the P-194 adopted an orthodox configuration.

Data File

Mission:	Attack, reconnaissance, dive bomber and heavy fighter
Final development status:	design
Engine:	one BMW 801 D piston engine + a BMW 003 or Jumo 004 turbojet
Maximum speed:	715 km/h (with auxiliary turbojet on)
Range:	1,070 km
Weight:	9,330 kg (with Jumo engine)
Span:	15.30 meters
Length:	11.80 meters
Armament:	two Mk 103 cannons, two MG 151/20 cannons and up to 1,000 kg of bombs

Luftwaffe - confidential - Fundamentals of modern aeronautical design

"T" tail

Focke Wulf Ta 183

Many authors consider the FW Ta 183 the most important of all German aeronautical projects designed during WWII. Its extremely advanced configuration is evidenced by the T tail, created by Hans Multhopp.

If it ever had been built and put into action, its very compact dimensions would probably have made it very agile and endowed it with an outstanding performance.

The first prototype would employ a Jumo 004 engine, since the definitive engine, the HeS 011 was still not available. As it happened to many other late war projects, the Ta 183 would feature mixed construction methods, using metal for the fuselage and wood for the wings. The Ta 183 could also be adapted for the fighter bomber role, carrying up to 500kg of bombs and could also be employed for reconnaissance.

Data File

Mission:	single place day fighter
Final development status:	wind tunnel models
Engine:	one HeS 011 turbojet engine
Maximum speed:	1,017 km/h
Range:	969 km
Weight:	4,291 kg
Span:	10.00 meters
Length:	9.20 meters
Armament:	four Mk 108 cannons

Butterfly tail

He P-1079

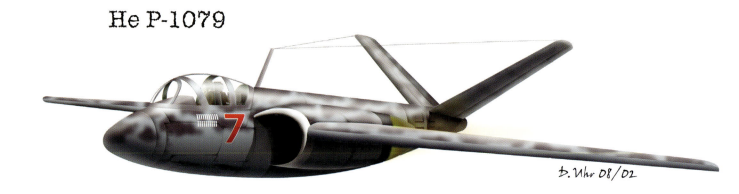

There were no fewer than five different versions of this twin engine night fighter. The hereby described version, known as P-1079A, was characterized by the prominent butterfly tail. The wings were swept at 35° and were slightly inclined downward. Jumo 004 or HeS 011 engines would be mounted close to the fuselage sides. Pilot and radar operator would seat, back to back, just as in the He 219. The aircraft would be armed with 30mm MK 108 cannons. An interception radar would be standard equipment.

Data File

Mission:	two place night fighter
Final development status:	design
Engine:	two HeS 011 or Jumo 004 turbojet engines
Maximum speed:	940 km/h
Range:	2,700 km
Weight:	10,000 kg (loaded)
Span:	12.70 meters
Length:	14.00 meters
Armament:	four Mk 108 cannons

VTOL

The Dornier Do 29 (first flown in 1958) had a configuration similar to project Fl 269. The propellers tilted down in order to perform short takeoffs and landings, just the same way as in the Focke-Achgelis fighter.
Photo: Dornier. Courtesy Franz Selinger.

This chapter investigates German research into vertical take-off and landing aircraft (VTOL).

Even before the beginning of World War II, Germany possessed solid experience in the development of vertical take-off aircraft, especially those equipped with rotating wings, better known as helicopters.

In 1923 the Spaniard Juan de la Cierva invented the autogiro. This can be characterized as a kind of hybrid aircraft, half way between airplane and helicopter. With this type of flying machine, the rotor is mounted on top of a pylon above a conventional fuselage but in contrast to a helicopter, the rotor blades turn freely and are not connected to an engine. Thrust is generated by another propeller driven by an engine on the fuselage nose. Forward movement causes the rotor to turn, creating a vertical lift force which allows the elimination of conventional wings.

Around 1930 Cierva toured Germany and Switzerland, making a series of demonstration flights. The interest generated prompted the Focke-Wulf company to acquire the rights to manufacture Cierva's autogiros. These, the C19 type, were powered by German engines and stimulated the design and creation of a German autogiro .

This, the Focke-Wulf Fw 186, was built from the modified fuselage of a Focke-Wulf Fw 56 *Stösser*, and was entered in the design competition for an observation aircraft for the Luftwaffe. Other entries included the Siebel Si 201 and the Fieseler 156 Fi *Storch*, the eventual winner.

In succeeding years, helicopter development and manufacture in Germany was concentrated in the hands of two companies: Flettner and Focke-Achgelis, each proposing different propulsive/rotor systems.

Focke-Achgelis, founded by Henrich Focke, (who years earlier was one of the original founders of Focke-Wulf) adopted the use of twin rotors, located on either side of the fuselage, to eliminate the torque generated by a single rotor.

Flettner, founded by Anton Flettner, used a system of intermeshing rotors, to achieve the same results.

The two companies developed a series of successful aircraft culminating with the transport helicopter Focke-Achgelis Fa 223 *Drache* and the Flettner 282 Fl *Kolibri*, for which a variety of missions were foreseen, mainly as shipborne anti-submarine patrol aircraft.

These rotorcraft had only a small part in Germany's war effort since they were produced in very small numbers. This was partly because the factories where they were built became priority targets for the allied bombing campaign. This indicates that rotorcraft's great military potential was recognized, as it proved later during the Korean and Vietnam wars.

At first, only non-offensive missions were foreseen for this type of aircraft. They were especially adapted for troop and equipment transport missions to places difficult of access (as for the Fa 223) or as patrol and reconnaissance aircraft (Fl 282).

With the turn of the tide in favor of the Allied armies, it became even more important to develop aircraft for the defense of Germany. This was even more vital as Luftwaffe bases became constant targets for American and English attack planes.

This situation stimulated the imagination of German designers, always ready to present unusual alternatives and solutions to complex problems.

The time was ripe for a type of aircraft that could take off from extremely confined spaces, climb straight up quickly to the altitude of the attacking aircraft to intercept them, returning to base without need of a long runway.

The first step towards a target defense interceptor was the Messerschmitt Me 163 *Komet*, the rocket fighter originally conceived by Alexander Lippisch. Its extremely limited radius of action determined that it must be based on airfields next to potential targets, such as the plants where synthetic fuel was manufactured for Luftwaffe aircraft. This fuel was a strategic resource which became increas-

A Cierva C30 autogiro preserved at the *Museo Nacional de Aeronáutica* in Buenos Aires.
Photo: Claudio Farias.

① Start

② Climb

③ Attack

④ Dive

⑤ Cockpit separation

⑥ Fuselage and engine recovery

⑦ Pilot landing

② The Me 163 took off as bombers came within 42 km from the airfield.

③ The *Komet* had an incredible climb rate (81 meters/second) and reached 8.000 meters in approximately two minutes and a half.

④ Climb angle: 45º

⑤ With its fuel depleted (after 7,5 minutes), the *Komet* would return to base gliding.

⑥ The *Komet's* range was approximately 35km.

ingly scarce and decisive for the continuation of the war.

The *Komet* still needed long runways for take-off and landing. Burn marks left on the soil by the Walter engines were easily identified by photographic reconnaissance and the sites were promptly attacked by allied forces.

What the Luftwaffe needed was an aircraft that did not require an airfield for take-off, in other words, one that could take off vertically. This way the bases could be installed in forests or other environments where they could be well hidden.

The Bachen Ba 349 *Natter* was a project bred from this necessity, but it is better defined as a manned missile rather than an airplane. This very small aircraft took off vertically from a wooden and metal supporting structure. Coming close to the fleet of attacking bombers, the pilot would choose a target and fire a battery of unguided rockets (something akin to the effect of firing a shotgun). He would then immediately abandon the *Natter* using his parachute. The fuselage would be recovered by way of a parachute installed internally, allowing the precious Walter rocket engine to be re-used. The rest of the *Natter* was completely expendable.

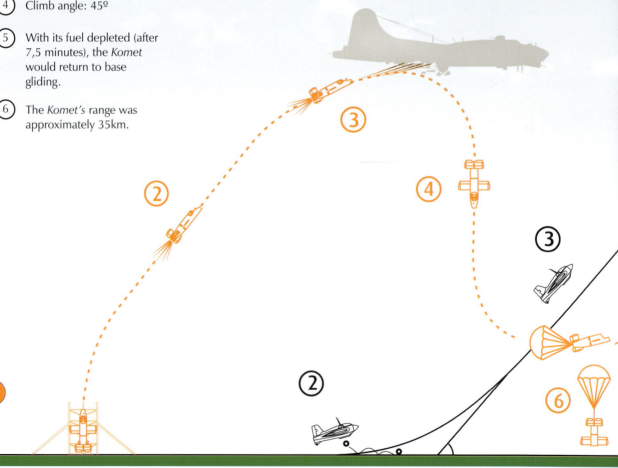

Other manufacturers tried to develop projects combining the characteristics of vertical take-off and landing with the operational requirements of conventional combat aircraft. These projects can be divided in two basic categories: convertiplanes and tail-sitting aircraft (coleopters).

CONVERTIPLANES

The first German design to adopt the convertiplane configuration was the Weser WP 1003/1, originally conceived in 1938 by Dr. Adolf Rohrbach. It was intended for high speed reconnaissance missions and civil use (this is not surprising, since the aircraft was designed before the war).

The WP 1003/1 can best be described as a tilt-wing aircraft. A pair of rotors would be mounted at the tips of each wing's outboard section. These wing segments would turn approximately 90° up to effect vertical take-offs and landings.

After reaching some altitude, the wings would be turned horizontal to fly like a conventional aircraft, with a predicted maximum speed of 650 km/h thanks to a DB 600 piston engine in the fuselage.

Although this project went only as far as preliminary drawings, it anticipated the arrangements found on some post-war convertiplanes, like the experimental Vought XC-142A, which flew in 1964.

Around 1940, in response to a RLM contract, Focke-Achgelis initiated the design of the convertiplane Fa 269. This was the creation of Henrich Focke and engineer Paul Klages, partly inspired by a Dornier company patent.

This type of aircraft was intended to combine vertical take-off and landing with high horizontal speed by means of a metamorphosis in flight.

Perhaps the most interesting aspect of the Fa 269 was its propulsion system: a BMW 132K piston engine mounted inside the fuselage. The engine would transmit the power through a complex system of axles and gear boxes to a pair of pusher propellers behind the trailing edge of the wings.

With the propellers facing backwards, the Fa 269 would fly normally, like other aircraft that had adopted this pusher configuration (for example the

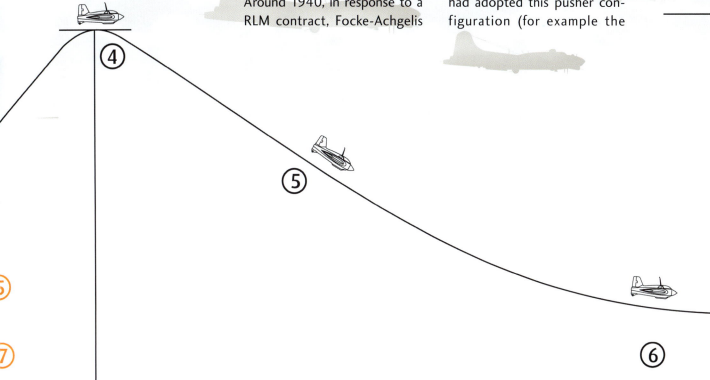

Luftwaffe - confidential - Fundamentals of modern aeronautical design

Artist's impression showing the *Convertiplano*, flying with a Gloster Meteor as chase plane. Illustration: Daniel Uhr.

Bell *Airacuda*). Maximum speed would be around 600 km/h, quite adequate for a bomber destroyer.

To take off or land, the complete set of propellers and transmission axles would be tilted down, to an angle of 85^0, so that the propellers would be turning in an almost perpendicular plane relative to the fuselage of the aircraft.

In order to provide the necessary clearance for the propellers to turn freely without touching the ground, the Fa 269 would be provided with a very long and complex telescopic landing gear which would extend from compartments located in the fuselage.

The project was evaluated through wind tunnel models and in 1942 a mock-up was under construction. It was destroyed during a bombing attack on the Focke-Achgelis plant in Hoykenkamp.

The Fa 269 project was cancelled shortly afterwards, but the concept returned to Germany in 1958, by way of the experimental Dornier Do 29. This could better be described as a STOL (short take-off and landing) aircraft and rather than a pure VTOL. The pusher propellers could be tilted down in order to shorten the take-off and landing run. Three prototypes, powered by Lycoming engines, were built.

After the end of the war Henrich Focke, together with other Focke-Achgelis specialists, worked for some time in France and later in Holland, where in 1951 he initiated the development of another convertiplane aircraft, the *Heliconair*.

It was then that the *Centro Técnico de Aeronáutica* was created so that Brazil could search for projects and foreign specialists in order to stimulate aeronautical development in the country. The CTA, Center for Technical Aeronautics, (today known as *Comando-Geral de Tecnologia Aeroespacial*), is located in São José dos Campos in the State of São Paulo.

Professor Focke was invited by Colonel Aldo Weber Vieira da Rosa to come to Brazil to get to know the facilities at ITA (*Instituto Tecnológico Aeronáutico* – Aeronautical Technological Institute) and of the CTA. Focke agreed to move to Brazil intending to help develop the *Heliconair*, now renamed *Projeto Convertiplano*.

A wind tunnel model of the *Convertiplano* designed by Professor Focke at the CTA. Photo: courtesy CTA.

Below - Brazil belongs to the group of nations where German aeronautical influence was most felt. The image shows the special test rig built to test the rotors and transmission of project *Convertiplano*. Photo: courtesy CTA.

Focke's team consisted of some 20 European engineers. Among them was Hans Swoboda, a former Focke-Wulf employee.

Initially, the *Convertiplano* would be a 'proof of concept' aircraft, used to evaluate different flight regimes (horizontal, vertical and transition). Its construction would have to be achieved with the scarce available resources.

Focke intended to build this test aircraft by adapting a *Spitfire* fighter bought in Europe. He would modify its fuselage in order to install a jet engine as well as the transmission axles and tilting propellers.

The initial design envisaged the installation of a British turbo-prop Armstrong Siddeley *Double Mamba* engine. Perhaps alarmed by the disastrous experience of the donation of a Rolls-Royce *Nene* engine to the Soviets (copied and installed in the deadly MiG 15), the sale of the *Double Mamba* was not authorized by the British government.

Focke had no option but to modify the project in order to accommodate a gigantic Wright R-3350 piston engine, as used in the Lockheed Constellation airliner.

The *Constellation's* huge radial engine would never fit inside the slender ballerina-like fuselage of the *Spitfire*. This forced the creation of an entirely new fuselage, as well as a redesigned empennage, featuring a 'T tail' (*a la* Multhopp).

Prototype construction was preceded by the construction of a group of test rigs where the propeller and transmission system was exhaustively tested for about 3,000 hours.

The project moved slowly on, facing all kind of difficulties, a direct consequence of the nature and complexity of the problems arising.

Brazil, still in the infancy of its aeronautical industry develop-

Luftwaffe - confidential - Fundamentals of modern aeronautical design

116

The configuration of the *Weser* WP-1003 would make it specially suited to operate from the German aircraft carrier *Graf Zeppelin,* had it been completed and put into operation. Illustration: Daniel Uhr.

ment, tackled an extremely ambitious project, as ambitious as projects developed later by much more advanced nations endowed with an abundance of resources, about which the Brazilian and German designers could only dream.

Projects similar to the *Convertiplano* were in development in other parts of the world at approximately the same time Brazil was trying to build a prototype.

In the USA, Curtiss-Wright developed an aircraft with a very similar configuration, the Curtiss X-19. As with the *Convertiplano*, the X-19 had four tilting propellers mounted at the tips of a set of tandem wings. It performed its first flight in June 1964.

Bell Aircraft Corporation, one of the USA's most experienced companies in the development of helicopters and VTOL aircraft, in 1941 registered the patent, granted in 1945, for a coleopter fighter powered by a piston engine. In the following years, the company continued research into this concept, beginning in 1956 with the experimental project X-14.

Bell later developed the Bell X-22, which also used four sets of tilting propellers mounted in the wing tips, using the principle of the ducted fan. It flew in March 1966.

The Bell convertiplane family had continuity with the 1977 project XV-15. In this aircraft, the engines with propellers together were directly mounted at the tips of the wings, thus eliminating part of the complex transmission system. The entire power unit tilted to allow landings and vertical take-off.

Successful tests spurred the development of a larger aircraft, capable of performing a range of missions, not only for the military, but also for the civilian market. The outcome was the Bell XV-22 *Osprey*, which is used today by the US Marines.

From the above examples, it is evident the Brazilian CTA at that time embarked on a project that, in many ways, was years ahead of similar developments in countries with much more solid aeronautical traditions

As could be expected in such situations, the attempted step was longer than the legs. The *Convertiplano* project died due to a lack of funds and the insurmountable difficulties found in its development. The cost exceeded 8 million dollars.

Henrich Focke left Brazil in 1955 for the USA. Yet some members of his team decided to remain in Brazil and continue the development of a conventional helicopter, initiated under Focke's direction, the *Beija Flor* (Hummingbird).

The *Beija Flor* was built and tested to make its first flight in 1959, but after an accident that destroyed the prototype, it was abandoned, in 1965.

Henrich Focke was not the only expatriate helicopter designer who came to Brazil.

In 1950 the Austrian engineer Paul Baumgartl, invited by the *Fábrica do Galeão*, arrived in Rio de Janeiro. In the years preceding WWII the FDG was responsible for the construction of a series of Focke-Wulf aircraft for the Brazilian Navy. These included the Fw 44 *Stieglitz* and the Fw 58 *Weihe*. Plans to build the Fw 200 *Condor* patrol-bomber did not materialize, although two imported

Top - The Bell XV-3, first flown in 1955, was designed and built in response to a US Army/Air Force request for a VTOL research aircraft with two tilt rotors. Photo: courtesy NASA.

An MV-22 *Osprey* from Marine Medium Tiltrotor Squadron (VMM) 263 lands aboard the multi-purpose amphibious assault ship USS Iwo Jima. Photo: courtesy US Navy.

The images show the one and only prototype ot the Beija-Flor (Humming Bird) undergoing static tests and performing a test flight. The helicopter flew after Prof. Focke had already left Brazil and it's development was mainly the responsibility of engineer Swoboda. Photos courtesy CTA (Comando Geral de Tecnologia Aeroespacial)

Fw 200 *Condor* airliners were operated by *Sindicato Condor.*

Baumgartl made a name for himself during the war by inventing an intriguing strap-on helicopter to be worn by Wermacht soldiers. This machine, called *Heliofly I,* was tested in 1941. This was followed by the *Heliofly III/57,* another one-man helicopter worn directly on the soldier's back like a rucksack.

In Brazil, Baumgartl designed a series of helicopters and rotor-kites starting with the PB-61, flown first in 1952.

This experiment evolved into the PB-63, Baumgartl's most sophisticated machine. It was conceived as a multi-purpose light helicopter, similar in general arrangement to Bell's Model 47.

The prototype employed a structure of welded steel tubes and was powered by a Continental C-85-12 engine driving a two bladed plastic-covered wooden rotor.

The prototype flew for the first time in 1953 and performed many tests in the next six months, showing good results. Yet by that time, the Ministry of Aeronautics had leased the *Fábrica do Galeão* facilities to *Fokker Indústria Aeronáutica SA,* a company recently created with Dutch and Brazilian capital, with the purpose of building a series of Fokker S11 trainers.

From then on the fate of the PB-63 becomes obscure. Some sources claim the prototype reappeared in 1955 during an industrial fair at the *Ibirapuera Pavillion,* in São Paulo, as the latest project by Grassi, a traditional manufacturer of bus bodies. It is said that Grassi had plans to initiate series production of the PB-63, but this never happened.

A contact with the *Parque de Material Aeronáutico do Galeão,* as the former *FDG* is known today, suggests another version of the PB-63's final destiny, saying simply that the prototype was destroyed in an accident.

Baumgartl's last design in Brazil was the jet powered PB-64, better described as a gyrocopter. For propulsion it employed two pulse-jets mounted at the tips of a stabilizing beam, set across the rotor. These engines, each producing 13 kg thrust, were developed at the *Instituto Tecnológico de Aeronáutica.*

The development of pulse jets in Brazil can also be linked to the presence of a German engineer: Alexander Heinrich Frank, who worked at Fieseler during WWII perfecting the Fi 103 flying bomb pulse jet engine.

He arrived in Brazil in 1954, and around 1957 was put in charge of the engine department of the CTA. There, in conjunction with Brazilian engineers, he began the development of pulse-jets which were to be used as auxiliary engines for military aircraft.

The first prototype pulse-jet engines were about one meter long, 10 centimeters in diameter, and were statically tested

The Baumgartl PB-63 flying over Santos Dumont airport in Rio de Janeiro.
Illustration: Daniel Uhr.

in a purpose-built test stand. These test engines were followed in 1958 by a bigger one, 2.50 meters long and 23 centimeters in diameter, of which three prototypes were built.

These larger engines were designed to power what would be the first Brazilian prototype jet aircraft, the IPT-15 *Besouro* (Bug), designed at the IPT – *Instituto de Pesquisas Tecnológicas* (Technological Research Institute) in São Paulo. This experimental aircraft was to be a jet engine test bed, and was designed to take both pulse jets and small turbojet engines.

Although the prototype was finished (less engines) it made only a few towed flights, as the CTA decided to terminate the pulse jet program.

COLEOPTER AIRCRAFT

Confronted with the necessity to create a vertical take-off and landing combat aircraft, some German designers opted for an unorthodox solution: stand the aircraft on its tail, so that the fuselage nose pointed vertically upward. It could then take off using the thrust generated by a set of propellers. This type of aircraft is known as a Coleopter (a term used to identify insects such as the Lady-Bug).

Heinkel was one of the companies that pursued this concept and its designs used two piston engines to power a set of counter-rotating propellers, as a way to eliminate the torque forces.

The *coleopter* fighter Convair XFY-1 *Pogo*. Photo: courtesy US Navy.

Engineers Kurt Reiniger and Gerhard Schulz had begun work on a family of coleopters around 1944. They used Daimler Benz piston engines (DB 603 or DB 605) and a ring-shaped wing like the shroud that covers the propellers of a table fan.

A total of four variants received the generic denomination *Lerche* (Lark). It was foreseen that with two crew members, they would be used as both day and night fighters and for ground attack.

Artist's impression of the IPT-15 *Besouro* (Bug) illustration: Daniel Uhr

A final development in this line was the *Wespe* (Wasp). These were similar to the previous projects, but powered by a DB PTL 021 turbo prop engine, a conversion of the Heinkel HeS 011 turbojet, which also moved a set of counter-rotating propellers. No prototype was built, although preliminary wind tunnel tests of some components were made.

Of all of the VTOL aircraft projects developed in Germany, most remarkable was the Focke-Wulf *Triebflügel*. This was a direct inspiration for the coleopter fighters developed for the American Navy after the war.

The idea came from Erich von Holst, a professor of zoology at the University of Göttingen.

He created the term *Triebflügel* (thrust wing), based on years of studying the flight of insects and birds. His research was particularly based on the observation and analysis of flight of the dragonfly, an insect that can fly in all directions (forward, backwards, laterally and vertically) at high speed, and can also hover.

Erich von Holst believed that the principles of animal flight could be applied to flying machines created by man.

In order to investigate this he created and patented a series of model aircraft that reproduced the characteristics of bird flight. These models spanned from 0.97 to 2.45 meters, and used an ingenious mechanism to reproduce the beating of birds' wings.

The application of these principles in the construction of aircraft was not feasible so von Holst turned his attention toward the dragonfly, whose wings have a different drive.

Together with two aerodynamics specialists at the AVA Göttingen, Professors Dietrich Küchemann and K. Wolf, von Holst concluded that the application of the principles of flight of the dragonfly to an aircraft should consider first the transformation of wing beating into a rotating movement.

To achieve this, the elements that generate lift (wings) and thrust (propellers), would have to be integrated in a single mechanism.

To test and demonstrate the viability of his ideas, von Holst initiated the construction of a series of models powered by rubber bands. These were presented during a 1940 scale model competition in the city of Breslau.

The discoveries derived from these tests were published in 1942, under the title *Die Triebflügel*, in the *Jahrbuch*

General views of the Heinkel *Wespe*. Illustration Daniel Uhr.

General views of the Heinkel *Lerche*. Illustration Daniel Uhr.

der Luftfahrtforschung (Yearbook of Aeronautical Research). This publication attracted the attention of specialists at AVA Göttingen, who then conducted wind tunnel tests with von Holst's models.

These events coincided with the increasing interest of the Luftwaffe in using vertical take-off aircraft.

The ideas of Von Holst, perfected by Hans Multhopp (see chapter two), were incorporated into the development of a revolutionary VTOL fighter, known as *Triebflügeljäger*.

Focke-Wulf's design added another bold element to von Holst's already unusual concept: propulsion would come from three small ramjets mounted in the tips of the rotor's propeller blades.

The propellers would be mounted in the central part of the fuselage and would turn freely, driven by the thrust of the ramjet engines, thus at least theoretically eliminating the torque effect.

The small ramjets were the creation of Professor Otto Pabst, a specialist in gas dynamics who developed a ramjet of compact dimensions, in both length and diameter, although their thrust was greater than that of conventional examples.

Models of these ramjets were wind tunnel tested. At a later stage, it was intended to perform live flight tests by installing two of them in the wing tips of a Focke-Wulf Fw 190.

The design of the *Triebeflügel* went as far as the making of detailed drawings but no prototype was built.

Even so, the concept seemed promising enough to interest the Allied nations who began their own coleopter aircraft programs after the end of the war.

SNECMA's ad promoting their strange *Coleopter* fighter. The illustration suggests that the aircraft could also be employed as a naval fighter, taking off from platforms mounted in small ships.
Photo: Aviation Magazine.

In France, German engineer Helmut von Zborowski, who was a rocket specialist at BMW during the war, was responsible for the research into a VTOL aircraft manufactured by SNECMA, the aptly named C 450-03 *Coléoptère*.

The *Coléoptère* was fully jet powered, using an ATAR 101EV engine (the ATAR engines were a development of the BMW 003 jet). As with the Heinkel projects, the *Coléoptère* also used a ring-like wing. Its first flight occurred in 1959, but the prototype was destroyed, happily without causing injury to the pilot, and the project was cancelled.

The similarity between the *Coléoptère* and the Heinkel projects was the cause of a long judicial dispute ending with the recognition, by the German Patent Office, of the precedence of Reininger's designs.

The coleopter concept also found fertile ground in the USA, especially within the US Navy. The Navy saw in vertical take-off aircraft the perfect solution for the problem of providing escort aircraft for convoys and warships.

By 1947, the US Air Force and the US Navy had sponsored a series of conceptual studies of VTOL aircraft, known generically as the *Hummingbird Project*.

Later the Navy continued this line of inquiry using not only the results of the *Hummingbird Project*, but also captured German documents related to the *Triebflügel* project.

At the beginning of the 1950's, the result was that Convair and Lockheed began designing two coleopter aircraft, powered by turbo prop engines. These were intended to operate as carrier aircraft for the US Navy.

The first one to fly was the Lockheed XFV-1 *Salmon*, in March 1954. Its initial test flights were conducted from conventional takes-offs (in the horizontal mode), using a complex landing gear bolted directly to the fuselage and wings of the airplane.

All flight tests performed in the vertical mode were conducted at altitude and the airplane never proved its ability to take off and land from the vertical position.

The Convair airplane, the XFY-1 *Pogo*, made its inaugural flight in August 1954. Compared with its Lockheed competitor, the *Pogo* could take off from a vertical position and land in the same way.

The flight tests, however, pointed to an insurmountable problem: only very experienced pilots were capable of putting the aircraft down 'in reverse gear,' and even then, only under ideal flight conditions.

To expect that an inexperienced pilot would be capable to carry out such exploits on the deck of a ship pitching in heavy seas would be unreasonable. This sealed the fate of both the *Salmon* and *Pogo* and it appeared that the possibility of creating a functional VTOL aircraft would be abandoned.

The *Coleopter* concept proved to be a dead-end. The dream of a functional pure VTOL aircraft would become reality only with the advent of a tiny British airplane called *Harrier:* whose name and subsequently distinguished history dispenses with formal introductions.

During the Cold War years, a novel approach to the problem of creating a target defense interceptor, and at the same time, dispensing with airfields was the so called Zero-length launch system.

The idea was conceived in the United States and also tested by the Soviets, and consisted in launching a fighter aircraft, initially a Republic F-84 *Thunderjet* and later a North

A F-100 *Super Sabre* takes off using the Zero-length launch system.
Photo: courtesy USAF.

American F-100 *Super Sabre*, using huge rockets attached to the aircraft.

The aircraft would be mounted on an inclined platform and the JATO (Jet Assisted Take Off) rocket would be fired, rapidly accelerating the fighter. Once flying speed was reached the booster rocket would be jettisoned and the aircraft would fly to its target.

The concept was extensively and successfully tested but never became operational.

To close this chapter, mention must be made of another intriguing project attributed to Professor Henrich Focke.

A 1944 patent attributed to Focke describes a discoid aircraft (similar in planform to Charles Zimmerman's *Flying Pancake)*, but featuring a set of counter-rotating propellers mounted inside the circular fuselage and powered by a non-specified jet engine.

It is not known what kind of mission that this aircraft would perform and it was only after the war that a 1/10th scale wind tunnel model was built in Bremen to investigate its flight characteristics. Professor Henrich Focke also filed a new patent in 1957.

Artist's impression of the Focke-Wulf VTOL Project. Illustration: Daniel Uhr.

Convertiplanes

Fa 269

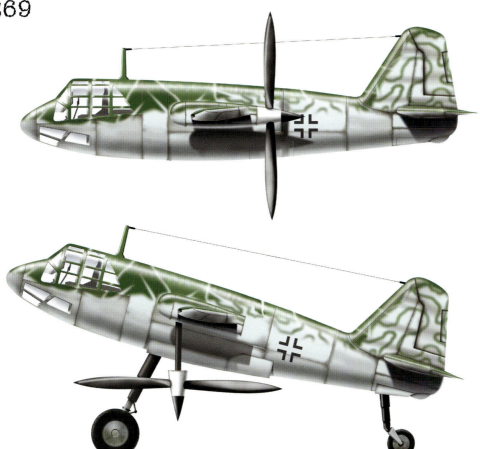

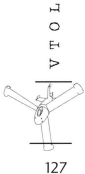

VTOL

Because of the bombing and destruction of the Focke-Achgelis facilities, little information survived on project Fa 269. Some data was listed in the British Intelligence Service's report "German Aircraft New and Projected Types," compiled by F/Lt J.L. Newton, in 1946.

Data File

Mission:	single place day fighter
Final development status:	Mock-up
Engine:	one BMW 132K engine
Maximum speed:	600 km/h
Range:	---------
Span:	10.00 meters
Length:	8.90 meters
Height	3.20 meters
Armament:	two Mg 151/20 cannons and one MK 103 cannon

Luftwaffe - confidential - Fundamentals of modern aeronautical design

Coleopter Aircraft

Focke-Wulf Triebflügel

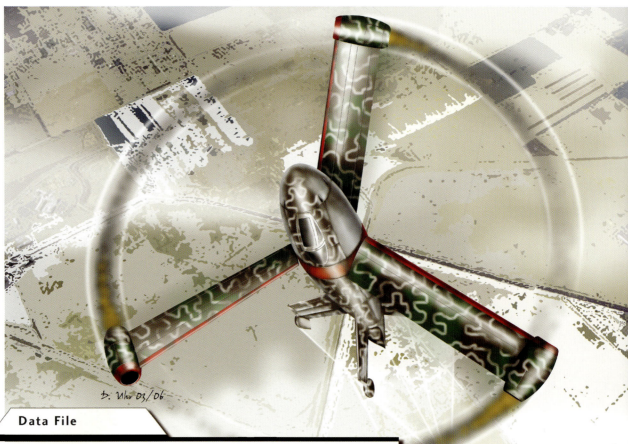

Data File

Mission:	single place day fighter
Final development status:	design
Engine:	Three Pabst ramjets (840 kg thrust approximately) plus three Walter rocket engines (300 kg thrust), to accelerate the ramjets to operating speeds
Maximum speed:	840 km/h (at 14,000 meters)
Range:	2,400 km
Rotor diameter:	10.75 meters
Length:	9.35 meters
Weight:	5,175 kg
Armament:	two Mg 151/20 cannons and two MK 103 cannons

The *Triebflügel* would be built entirely of metal and would take off and land from the vertical position, supported on its cruciform tail. At the tip of the four tail stabilizers a small support wheel would be installed and the weight of the aircraft would be mainly supported by a larger central wheel installed at the rear extremity of the fuselage. Given its peculiar configuration, further consideration should have been given to the problem of how the pilot would abandon the aircraft in case of an emergency. Besides the installation of an ejector seat, it would probably be necessary to include a system to blast off the rotor blades. After the war, many engineers voiced their opinion that the *Triebflügel* would never be capable of taking off, flying, and landing in a satisfactory form.

Special Missions

Silbervogel's wind tunnel model.
Photo: courtesy Franz Selinger.

In this chapter are gathered projects that sometimes defy a more precise classification. Some of these ideas had no continuation after 1945, but others were decisive in the development of military aviation during the Cold War.

Special attention is given to aircraft that were powered by rocket engines. This new kind of propulsion, used with relative success in the Messerschmitt Me 163 *Komet*, would make possible the development of aircraft whose operational profile could provide substantial advantages in combat. This was the case with the DFS 228, a spy airplane whose operational ceiling and speed would make it virtually invulnerable.

The rocket engine's enormous power would also allow German scientists to initiate the development of an aircraft that was intended to explore flight at supersonic speeds. This, the DFS 346, began construction in the workshops of the Siebel company. It reached prototype testing only in the Soviet Union, under the Russian authorities.

Rocket propulsion was also employed in the *Silbervogel* project. This was to be an intercontinental bomber capable of flying in the upper limits of the atmosphere and of reaching hypersonic speeds. In certain aspects, it anticipated the operational profile of NASA'S Space Shuttle.

The chapter describes the minimum and maximum dimensional limits of German aeronautical projects, minuscule parasite airplanes and miniature fighters, and at the other

Artist's impression of the Fieseler Fi 333. Illustration: Daniel Uhr.

extreme the gigantic Daimler-Benz combination aircraft.

THE PACKPLANE

This idea was initiated by the Fieseler Company in 1942, and presented an entirely novel approach to the design of transport aircraft.

The Fieseler Fi 333 was a multi-purpose transport aircraft that would carry its cargo in a detachable pod beneath the fuselage. Upon arriving at its destination, the pod would be quickly removed and replaced either by an empty one or one carrying passengers instead of cargo. Pods could be configured for MEDEVAC (air ambulance) missions, substantially reducing the time required to load and unload

The Fi 333 also featured a fixed undercarriage mounted on very long legs that would provide adequate clearance for the installation and removal of the pods. The main landing gear had tandem wheels, which would improve the plane's ability to operate on rough terrain. This arrangement was previously tested on a Fieseler Fi 156 E-0 *Storch* (GG+XT).

With the pod removed, the tall undercarriage would permit the Fi 333 to transport complete aircraft fuselages and wings suspended under its own.

The Fi 333 remained in the design stage and only a partial mock-up was ever built, but the idea was resurrected in 1950 in the form of the Fairchild XC-120 *Packplane*. This was adapted from a standard C-119 *Boxcar* and featured the detachable pod concept proposed earlier by Fieseler. The prototype was thoroughly evaluated but the concept was not adopted.

The Fairchild XC-120 *Packplane* Photo: Fairchild. Photographer: Dan Frankforter.

The parasite fighter XF-85 Goblin. Photo: Courtesy USAF.

The XF-85 fixed to Monstro's trapeze during one of the docking tests. Photo: Courtesy USAF.

Artist's impression of the parasite fighter Messerschmitt P-1073. Illustration: Daniel Uhr.

PARASITE AIRCRAFT

The idea of an aerial aircraft carrier is not new; it can be traced back to WWI when the German company Siemens-Schukert developed an aerial torpedo glider that was to be launched from Zeppelins.

The idea gained a new impetus in the postwar period, when experiments were made in the United States with the launch and recovery of escort aircraft from US Navy airships.

The airships *Akron* and *Macon* were provided with internal hangars and mechanisms for launching and recovery of up to five Curtiss F9C-2 *Sparrowhawk* fighters. The first successful tests occurred in 1932.

The next step towards a flying aircraft carrier was taken in the Soviet Union. This involved the transportation of fighters carried both above and below the wings of gigantic Tupolev TB-3 bombers.

This project was initiated around 1931 and in its final incarnation foresaw the installation of up to five fighters, two Polikarpov I-5 biplanes above the wings and two Polikarpov I-16s below. In addition, a Grigorovich I-Z would connect itself to the bomber after take-off by way of a trapeze lowered from the fuselage of the mother-ship.

These so called *Zveno* combinations were used in combat during WW II as long-range bombers during the initial stages of the German invasion of the Soviet Union.

In Germany, the idea of carrying escort fighters inside strategic bombers found favor in the Messerschmitt company. In 1940, Messerschmitt initiated the design of a long range reconnaissance and bomber aircraft known as Project P-1073B. It was conceived specifically to attack enemy convoys over the North Atlantic. Powered by eight engines it would carry up to three escort fighters, known as P-1073A *Bordjäger* (onboard fighters), internally. Their folding wings would facilitate their loading inside the mother-ship.

Once launched, there would be no way for the escort fighter to return to the launching airplane. Each flight of a P-1073A would be potentially suicidal.

This idea was taken over and perfected in the United States by the McDonnell Company who, in 1947, designed the XF-85 *Goblin*. This parasite fighter that would be carried inside the huge bomb bay of the Convair B-36 bomber.

In contrast to the German project, the *Goblin* could be retrieved by the B-36 using the launching trapeze. This would be lowered from the bomb bay to launch and retrieve the *Goblin* once the mission was finished. The trapeze system was installed in a modified B-29 bomber (aptly nicknamed *Monstro*) and tested in 1948.

Tests demonstrated that the launching operation could be accomplished with relative ease, but the recovery of the *Goblin* was a much more difficult affair. The returning fighter would have to creep cautiously close to the launching aircraft, and would be submitted to the intense turbulence generated by the air displacement of the giant bomber.

In the first mooring attempt, the *Goblin* collided with the trapeze, damaging the cockpit canopy and forcing the pilot to make an emergency landing in the desert.

Later some successful mooring tests were made but the USAF lost interest. It became more interested in the development of long range escort fighters and in-flight refueling as a means of extending the range of aircraft.

SPY AIRCRAFT

Intelligence gathering was one of the first missions allocated to military aircraft, beginning with observation balloons in the 19th century. During WWI, balloons were again used as observation platforms.

The first heavier than air craft also began venturing over enemy lines, searching for vital information about movements and the location of the enemy's armies.

Initially, these flights were performed by unarmed aircraft, usually manned by a pilot and an observer who operated the photographic equipment. It was not uncommon for enemy aircraft to cross paths in the air while going to or coming back from their missions.

Very early, belligerent nations realized that action should be taken to hinder the free access of these spies to their airspace, denying them the precious information so easily obtained until then. The perfecting of antiaircraft artillery developed as a direct consequence.

Fighter aircraft also owe their existence to the same cause since their initial mission was the destruction of the enemy's observation/spy aircraft, as well as escorting and defending their own.

The experience acquired during the Great War made it clear that future reconnaissance aircraft would have to be specially designed or adapted for their missions. They would succeed only if they could remain hidden from, or out of reach of, antiaircraft artillery and enemy fighters.

To achieve these goals, two requirements were taken as the main parameters guiding the design of efficient reconnaissance aircraft: high speed and the ability to reach great altitudes.

In Germany, research on stratospheric aircraft had begun before WWII and gave birth to a series of experimental aircraft capable of carrying out missions above 10,000 meters altitude.

One of these was the Junkers Ju 86R, an adaptation of the obsolete Ju 86 bomber developed before 1939. Although this was successful in its initial missions, the British were quickly able to create an effective countermeasure: the *Spitfire* Mark IX fighter. This modified *Spitfire* was stripped of some of its armament and much of its armour to diminish the weight. The Mark IX did not have a pressurized cockpit which would have increased the weight.

This special *Spitfire* could reach the operational altitude of the Ju 86R and intercept it.

The Luftwaffe realized that it needed an aircraft whose speed and operational ceiling were well beyond that of the enemy fighters. Reaction engines (jets or rockets) opened the way to the development of a new family of spy aircraft virtually immune to interception.

The chance was taken with the initial deployment of the Arado Ar 234 *Blitz* (Lightning), the first jet powered reconnaissance aircraft.

Even while the Arado 234 was in its initial developmental stages, thought was already being given to future replacements whose performance would be even better.

One such project was the DFS 228, designed at the *Deutsche Forschungsanstalt für Segelflug* (DFS, German Research Institute for Gliding Flight) by Felix Kracht. It was derived from a series of studies investigating the possibility of a stratospheric glider with a pressurized cabin.

This led to the design of the DFS 54 which, although never built, proved the viability of building a small, efficient pressurized cabin which would accommodate the pilot in a prone position.

The DFS 228 was very similar to the DFS 54. Its basic configuration resembled that of a high performance sailplane, with long and thin wings specially adapted for flight at great altitudes. The DFS 228 could well be described as a powered glider since it would perform a great part of its mission with the engine shut off.

A specially adapted Dornier Do 217 bomber would be used initially to carry the DFS 228 to an altitude of 10,000 meters in a *Mistel* fashion (see COMBINATION AIRCRAFT).

After separation, the DFS 228 would fire a Walter rocket engine, similar to the one used in the Me 163 *Komet,* to climb up to about 22,500 meters. At this altitude the engine would

This rare photograph shows the DFS 228 preparing to land at Hörsching, piloted by Dipl. Ing. Ziegler, in November 1944. Photo: Dr. Manfred Reinhardt. Courtesy Franz Selinger.

The Lockheed U2 spyplane. Photo: courtesy NASA.

The Bell X-1 was the first airplane to break the so-called "sound barrier". Photo: courtesy NASA.

be shut off and the airplane would glide toward the mission's objective.

At regular intervals the engine would be fired and the DFS 228 would climb back to its maximum operational ceiling. It was calculated that the fuel load would allow this flight pattern to be extended for up to 45 minutes. With the fuel depleted, the airplane would return to base with a 1050 km gliding flight.

Since at operational altitude the installation of a conventional ejection seat was out of question, the DFS 228 made use of an interesting escape system to ensure the survival of the pilot in case of emergency. If forced to abandon the airplane the pilot would push a button to detonate a set of explosive bolts. This would separate the entire pressurized cabin from the fuselage. Moments after separation, a parachute would be deployed stabilizing the cabin and diminishing its rate of descent until it reached an altitude where the atmospheric pressure and the amount of oxygen present in air would allow the pilot to abandon the escape capsule safely.

At this the pilot would turn a lever to separate the front part of the cabin and eject the horizontal couch upon which he was lying. This would then allow him to release his safety belt, abandon the couch and deploy his personal parachute.

This ingenious capsule escape system was originally designed for the Heinkel He 176, the first airplane powered by a liquid fuel rocket (its first flight took place on June 20th, 1939). Tests were carried out using a Heinkel He 111 bomber to launch a cabin mock-up of a He 176, with a dummy pilot, to demonstrate that it was possible to abandon the escape capsule while it fell.

A prototype of the DFS 228 was completed and made about 40 test flights, launched from a Do 217K, but all tests were performed without rocket power.

This prototype was captured at the end of the war and taken to England, where it was carefully examined at Farnborough. Some sources mention that it was delivered to the British sailplane manufacturer

The Bell X-2 had swept wings and a detachable cockpit. Photo: courtesy NASA.

Slingsby Sailplanes Ltd., who developed the idea of a stratospheric glider, the Type 44, around 1955.

Unfortunately the prototype DFS 228 was not preserved, but its concept seems to have been incorporated into the design of the most famous spy aircraft of the 20th century, the Lockheed U-2.

The U-2, designed by Clarence 'Kelly' Johnson, adhered to the same principles as the DFS 228: to remain beyond the reach of its enemies by flight at extreme altitudes. As with the German project, the U-2 was capable of flying at more than 20,000 meters of altitude, allowing flight over the Soviet Union's territory with (apparent) total impunity during the Cold War.

The rapid development of air-to-ground missiles abruptly ended the glory years of the U-2 when the airplane piloted by Francis Gary Powers was downed by the shock wave from a barrage of Soviet missiles launched against it on May 7th, 1960.

As opposed to the DFS 228, the U-2 neither made use of an escape capsule, nor had it an ejection seat. The pilot used a pressurized suit. Powers had to jump out of the aircraft by his own effort.

Parallel to the DFS 228, the *Deutsche Forschungsanstalt für Segelflug* also explored another line of investigation to obtain knowledge and data about flight at supersonic speeds. This was a territory until then virtually unexplored by aeronautical research.

The DFS 346 was designed and conceived exclusively for this with no military applications. It was the result of years of theoretical inquiry at the aerodynamics research centers in Göttingen and Braunschweig, and its complexity meant it could not be built by the DFS workshops. Therefore, its construction was transferred to the Siebel company, manufacturer of the Siebel Si 204 transport and trainer aircraft.

The DFS 346 had wings swept at 45^0 and a 'T' tail like the Focke-Wulf Ta 183. It would be powered by two Walter rocket engines, mounted one above the other inside the fuselage.

As with its immediate predecessor the DFS 228, the 346 would make use of the

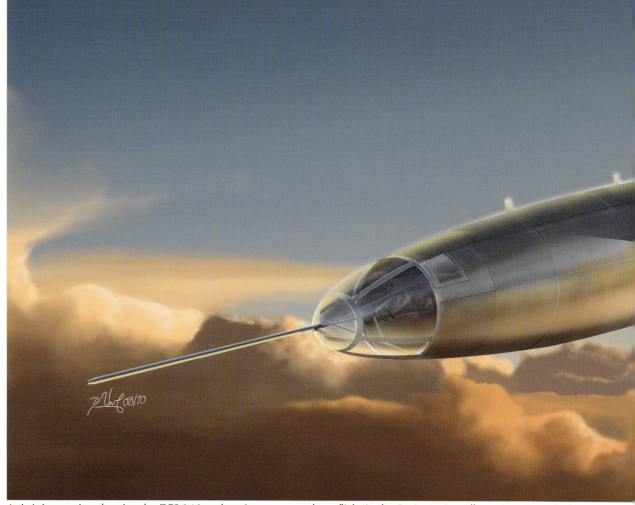

Artist's impression showing the DFS 346 undergoing a powered test flight in the Soviet Union. Illustration: Daniel Uhr.

capsule emergency escape system. (See above).

Project 346 was still at an early stage of development when the Siebel plant was captured by Russian troops. The Soviets showed great interest.

Some sources claim that a glider version of the DFS 346 (known as DFS 301) was almost ready when the Siebels factory was overrun. In October 1946, this prototype and all the material and personnel related to it, including the company test pilot, Wolfgang Ziese, were transferred to the area of Podberezhye, about 120 km from Moscow.

Two groups of German engineers had been established here. OKB 1 (Experimental Design Office 1) led by professor Brunolf Baade was responsible for the Soviet development of the Junkers Ju 287 and its derivative, the Ju EF 131 (see Chapter 1). OKB 2, headed by Hans Rössing, would continue the development of Project DFS 346, known now as *Samolyot 346*.

The glider version of the 346 began flight testing in 1948. It was launched from a Boeing B-29 bomber that had been seized during the war following a forced landing in Vladivostock while returning from a bombing mission to Japan.

During about 30 tests carried out with the glider, Ziese reached speeds of nearly 900 km/h, while diving from an altitude of 10,000 meters.

A series of tests were made to evaluate the functioning of the escape capsule. This time it was launched from a North American B-25 *Mitchell* bomber obtained from the USA by more orthodox ways, lend-lease.

Future pilots of the 346 would have to control it from the prone position, as with the DFS 228. For them to get used to this, two German gliders (known as *Liege Kranich* and *Liegebaby* – Prone *Crane* and

SPECIAL MISSIONS

Baby) were modified as instruction and training aircraft.

Tests with a powered version began in 1951, and in September of the same year it reached the speed of Mach 0.9, very close to breaking the sound barrier. The flight had to be interrupted when strong vibrations made the aircraft so unstable that Ziese lost control.

The German pilot initiated the escape system and the cabin detached from the fuselage at 6,500 meters. At 3,000 meters, the automatic system ejected Ziese and he reached the ground safely.

The victory over the so called 'sound barrier' would finally come in October 1947, at the hands of Chuck Yeager, piloting the experimental airplane Bell X-1. This was powered by a rocket engine, and as with the 346, was launched from a modified B-29 bomber. The main difference between the two projects was that the Bell X-1 had straight wings and no escape capsule.

The "X" series of experimental and research aircraft continued with the Bell X-2, designed to investigate flight at speeds up to Mach 3.5. In contrast to the X-1, it displayed wings swept at 40°. The X-2 would investigate for the first time the effects of heating generated by aerodynamic friction. For this reason, X-2 was the first aircraft to use exotic metallic alloys and materials in its structure.

At such extreme flight conditions, alternative and unorthodox escape methods for the pilot needed to be considered and Bell engineers adopted the same solution as in the DFS 346, a detachable escape capsule. The capsule was formed by the aircraft nose and cockpit, which would be separated from the fuselage by explosive bolts.

Unlike the German project, once he reached a safe altitude the pilot of Bell X-2 had to open the cockpit canopy and abandon the cabin unaided. This proved fatal in 1956.

After reaching the speed of Mach 3.19, pilot Milburn Apt lost control of the aircraft and was forced to use the cabin ejection system. The escape capsule detached from the fuselage and the stabilization parachute opened normally.

Later it was discovered that Apt had managed undo his safety belt and open the cockpit canopy but could not get out of the cabin. His body was found partly out of the cockpit wreckage.

In later years, escape capsules and jettisonable cabins have been used in aircraft such as the Sukhoi Su 17, a 1949 experimental prototype (not to be confused with the Su 17 fighter-bomber that NATO codename 'Fitter C'), and more recently in the General Dynamics F-111 variable sweep winged bomber.

The F-111 made its first flight in 1964. Its jettisonable escape pod was designed to allow the crew to remain inside it until reaching the ground. In case of a landing at sea, the module would float and guarantee the survival of the crew.

ROCKET AIRPLANES

The most ambitious project coming from German drawing boards during WWII was, without doubt, the sub-orbital bomber designed by Professor Eugen Sänger.

The origins of this idea can be traced back to 1932, when Sänger initiated the design of a *Stratosphären-Flugzeug* (Stratospheric Airplane), a civil aircraft.

Sänger also was one of the pioneers in research into liquid fuel rocket engines. In some respects, his work was more advanced than that undertaken by the group of Wernher von Braun and Walter Dornberger, in Peenemünde (the V2 missile project).

By 1936, Sänger had developed a high-pressure rocket engine, thus attracting the attention of the RLM (*Reichluftfahrtministerium* - Ministry of the Aviation of the Reich) concerning the military potential of his project. The RLM established a research center in Trauen and starting in 1939 a series of rocket engine tests was carried out, looking forward to the development of a rocket engine with 100,000 kg of thrust.

From this moment, the stratospheric airplane had its mission modified to that of a sub-orbital bomber, capable of carrying a military load to the United States, then gliding back to its base.

The design of the sub-orbital bomber was presented in a private report entitled *Über einen Raketenantrieb für Fernbomber* (On Rocket Propulsion for Long-range Bombers) and the aircraft received the codename *Silbervogel* (Silverbird).

The *Silbervogel* would be launched from a monorail about three kilometers long. The aircraft and engine would be mounted on a sleigh to run along the take-off track, accelerating to take off speed, driven by a booster rocket engine with 150,000 kg of thrust

Leaving the launching slope at the speed of Mach 1.5 it would adopt an angle of ascent of 30°, climbing until it reached about

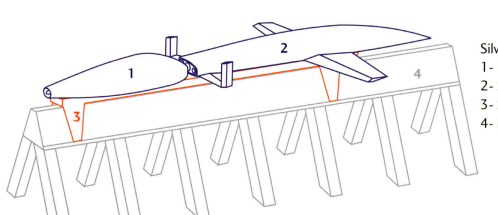

Silvervogel's takeoff system
1- Booster rocket
2- Sänger bomber
3- Sled
4- Monorail

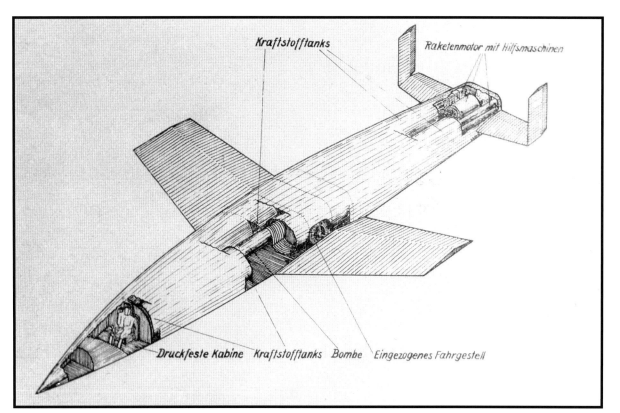

The illustration shows the arrangement of components inside the *Silbervogel*: *Druckfeste Kabine* (pressure cabin), *Kraftstofftanks* (fuel tanks), *Bombe* (bomb), *Eigezogenes fahrgestell* (landing gear), *Raketenmotor mit hilfsmachine* (rocket engine with auxiliary combustion chambers). Photo: courtesy Franz Selinger.

1,200 meters, when its own 100 ton thrust rocket engine would be fired.

The rocket engine would run for 4 to 8 minutes, consuming 90 tons of fuel, a mixture of combustible oil and liquid oxygen as oxidizer, taking the *Silbervogel* to an altitude of 145 km and a top speed of 22,000 km/h. This would place the *Silbervogel* in the realm of space flight.

Once the fuel was exhausted, the *Silbervogel* would commence a series of dives down to denser regions of the atmosphere, gaining momentum and speed to make an ascending arc at the end of each dive, like a roller-coaster. This ricochet effect, skipping over the denser parts of the atmosphere, is similar to the one obtained when a rock is thrown at a flat angle against the surface of a lake, making it bounce many times before finally sinking. By repeating this maneuver many times, it was calculated that a total range up to 23,500 km could be reached.

The *Silbervogel* project was abandoned when only some wind tunnel and friction tests had been made. One of the wind tunnel models is currently on exhibition at the Science Museum, in London.

The German sub-orbital bomber project did not escape the attention of the two superpowers, already flexing their muscles in preparation for the Cold War. Both the Americans and the Soviets showed a keen interest in Professor Sänger's design.

Legend says that the Soviet leader Joseph Stalin would have offered his weight in gold to whomever could disclose the whereabouts of Sänger and his wife, and assistant, Irene Bredt. Stalin even put his own son Vasili, in charge of the search for the German scientists.

During the 1950's, rumors circulated that the Russians were developing their own version of the *Silbervogel*. The press published illustrations that showed an aircraft of this kind marked with the red stars of the Soviet Air Force on the wings and tail.

In more recent years it was confirmed that a similar aircraft, powered by ramjets, was conceived by M. W. Keldysch, but it never reached the prototype stage.

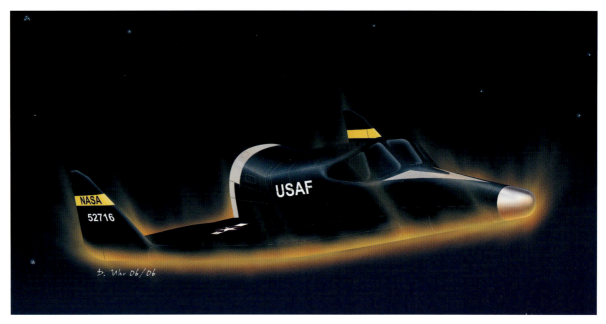

Artist's impression of the Boeing X-20 Dyna Soar. Illustration: Daniel Uhr.

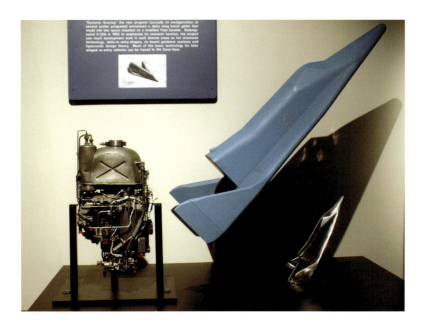

Wind tunnel models of the Boeing X-20 Dyna Soar preserved at the USAF Museum. Photo: Courtesy USAF.

In the United States, Sänger's ideas found fertile development ground after documents relating to the *Silbervogel* project were analyzed by American Intelligence.

The Bell Aircraft Corporation considered the project of a manned missile, known by the BoMi acronym (BomberMissile), and presented it at Wright Field in 1952. By 1954, the aircraft's mission had changed to strategic reconnaissance and other companies were invited to participate in the competition for its development. In 1956 the project once again included the option of a bomber aircraft, in addition to the reconnaissance version. Boeing, Convair, Douglas, McDonnell, North American, and Republic all took part in the competition for what was now known as RoBo (Rocket Bomber).

In 1957, the proposals presented by the competing companies were consolidated into a project called *Dyna-Soar* (Dynamic Soaring), now with the intention of developing a research aircraft to obtain data about flight at hypersonic speeds.

Boeing won the contract to develop the project, which gained the official designation Boeing X-20 *Dyna-Soar*.

In contrast to the *Silbervogel*, the X-20 would take off vertically, mounted on the tip of a

Titan IIIC rocket. After its mission in space it would return to the atmosphere gliding in the same manner as the later Space Shuttle.

The project evolved as far as a detailed mock-up but was abandoned in 1963, partly because of the successful missions of the first manned space capsule, project *Mercury*.

Its legacy was fundamental to the development of the Space Shuttle program, designed by Rockwell International.

However, the *Dyna-Soar* project was not the only legacy of the *Silbervogel* in the United-States. Professor Sänger's intriguing creation probably influenced another aspect of post-war American culture: the movies.

In 1951, Paramount studios released a science fiction film. *When Worlds Collide,* a George Pal Production starring Richard Derr and Barbara Rush, was directed by Rudolph Maté, and was the winner of a special effects Oscar.

In this early example of the 'Catastrophe Movie' genre, astronomers discover that Earth is on a collision course with the star *Bellus* and its planet called *Zyra*.

Scientists' calculations confirm that Earth will inevitably be destroyed by *Bellus* in the near future. This motivates a group of industrialists to undertake construction of a rocket designed to take 40 survivors of the human

In the Paramount Studios film "When World's Collide", the rocket that saves part of Earth's population takes off from a monorail track mounted on the base of a mountain and goes up to its summit. Illustration: Daniel Uhr.

Luftwaffe - confidential - Fundamentals of modern aeronautical design

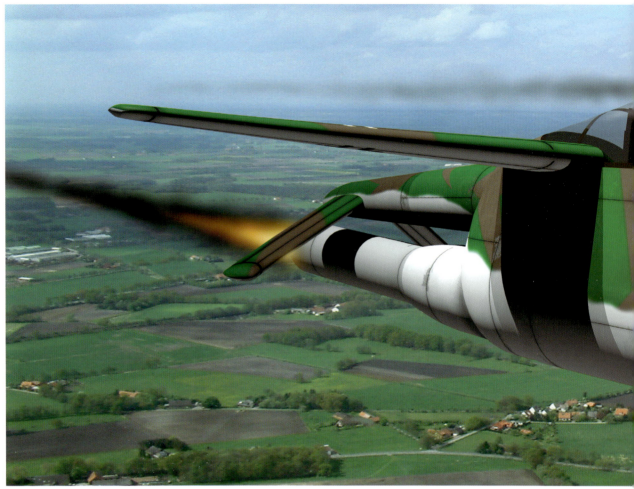

Artist's impression showing the BV 213. Illustration: Daniel Uhr.

race and animal pairs to planet *Zyra,* an obvious allusion to the Noah's Ark tale.

The spaceship, like a rocket but mounted in the horizontal position, is built in record time and takes off after accelerating on an extensive track that extends to the top of a hill.

Professor Eugen Sänger's opinion about the film is not known.

MINIATURE FIGHTERS

With the speedy deterioration of the military situation on all fronts, the Luftwaffe was forced to resort to desperate measures, searching for aircraft that could hold back the bomber armadas that were destroying German industrial plants and cities, day and night.

The German aeronautical industry was a primary target of the strategic bombing campaign. It was forced to work with an ever dwindling stock of raw materials and cope with the diminishing availability of specialized workers.

Jet aircraft, such as the Messerschmitt 262 and the Arado 234, proved to be only stopgaps against the continuing destruction of the Nazi military machine. Their manufacture involved the use of great amounts of increasingly scarce strategic raw materials. The jet engines and sophisticated systems also demanded great costs in workmanship.

Even desperate solutions such as the Heinkel He 162 *Volksjäger* (People's fighter), the aero-military version of the VW *Beetle*, were now considered excessively sophisticated considering the state of the German armament industry.

SPECIAL MISSIONS

The idea of the *Miniaturjäger* (miniature fighter) was to reduce much as possible the use of scarce materials such as aluminum, and to substitute steel or wood for them whenever possible.

To further save resources, the complex and expensive turbojet engines would be replaced by pulse-jets like those used on the Fieseler Fi 103 flying bomb, better known as the V-1. These consumed about 450 fewer man/hours in manufacture compared to turbojets and did not require sophisticated metallurgic materials.

Blohm und Voss, Arado and Heinkel presented designs for the *Miniaturjäger* competition which was announced in November of 1944.

The P-213 project, as could be expected from any design coming from Blohm und Voss, proposed very original solutions. It was based on concepts already used in the P-211, a competitor of the Heinkel He 162.

The concept of miniature fighters was completely abandoned in the years that followed the war as the rapid development of aeronautical technology saw the size and weight of fighter aircraft grow at an ever increasing pace.

Some such as the Lockheed F-104 *Starfighter* tried to reverse this trend but today the Lockheed-Martin F-35, for example, weighs about 22,680 kg loaded and is defined as a light fighter!

COMBINATION AIRCRAFT

The origins here are related to a series of experiments carried out by DFS to discover new methods for towing combat gliders, such as the DFS 230 and the Gotha Go 242.

Luftwaffe - confidential - Fundamentals of modern aeronautical design

The DFS 228 mounted over the Dornier 217 K-3 carrier plane (in *Mistel* fashion), preparing for its first flight in Hörsching during October 1944. In that occasion it was piloted by Dipl. Ing. Ziegler, seen standing in the aircraft's cockpit. Photo: DFS via Dr. Manfred Reihnhardt. Courtesy Franz Selinger.

These experiments became known by the codename *Mistel* and involved mounting a powered aircraft on a DFS 230 using a metallic support structure installed above the wings and fuselage of the glider. The combination would take off using the engine power of the airplane.

The first tests occurred in 1942 with a Klemm Kl 35 and DFS 230. Unfortunately, the Klemm's engine did not have enough power to take the combination off and required the assistance of a conventional towing airplane (a Junkers Ju 52).

In the next test the Kl 35 was replaced by a Focke-Wulf Fw 56 *Stösser*. This combination was capable of maintaining level flight once cruise altitude was reached but take-off still required the aid of a Ju 52.

The first successful flights were achieved with the installation of a Messerschmitt Bf 109E on the glider in 1943.

Later, in a proposal by Luftwaffe Captain Siegfried Holzbaur, the *Mistel* system was adapted to make airplanes into guided flying bombs.

The intention was to transform a Junkers Ju 88 bomber by removing its cockpit and replacing it with a 3,500 kg explosive charge. Above would be mounted a Bf 109 fighter to steer the bomber to the proximity of the target.

In the operational versions, the Bf 109 could be replaced by a Fw 190. The addition of auxiliary fuel tanks would allow an increase in its operational range.

On the way to the target, the Fw 190 pilot would control the Ju 88 through a sophisticated system of electric circuits (an early form of fly-by-wire technology). When close to the final destination, the pilot would aim the combination in the direction of the target and separate the fighter from the bomb. The bomb would then continue on its path until hitting the target as the Fw 190 returned to base.

This weapon was used in combat on the eastern front, attempting the destruction of bridges to halt the advance of Soviet troops. Yet in general, it proved to be a failure.

The idea of mounting one airplane above another was not entirely new. This has been mentioned above in the description of the aircraft parasites and the Soviet projects at the beginning of the 1930's.

Another important precedent

Luftwaffe - confidential - Fundamentals of modern aeronautical design

A formation of *Mistel* aircraft en route to their targets. Illustration: Daniel Uhr.

The Short-Mayo composite aircraft. Photo: courtesy of Aeroplane Photo Supply from the collection of Dan Shumaker.

SPECIAL MISSIONS

was the Short/Mayo project. Proposed in a 1932 memorandum by Major Robert Mayo, general manager of Imperial Airways, it involved launching a small aircraft from a larger airplane. The design of the smaller aircraft would be optimized for the crossing of the Atlantic Ocean.

In the final arrangement, operated in 1937, a four-engine S.20 Short seaplane called *Mercury*, was mounted above a flying boat, the S.21 *Maia*, derived from the S.23, *Empire* Class airliner.

The primary goal was to speed up the delivery of transatlantic mail between England and the United States. The combination would take-off joined and when approaching a pre-determined point, the S.20 separated from the launching aircraft and continued its flight to the USA. The system was put into use in July 1938.

The idea of using a carrier aircraft as a way to increase the range of a smaller aircraft gained new life at Daimler-Benz in the last months of WWII. The project was known as the Daimler-Benz *Projekt A*.

This was developed to create a weapon system capable of

Luftwaffe - confidential - Fundamentals of modern aeronautical design

The operational profile of the B-36/RF-84 composite aircraft was similar to that proposed for the Messerschmitt P-1073.
Convair ad published in Time Magazine, 1954.

The FICON concept under test. Photo: Courtesy USAF.

reaching the east coast of the US or strategic targets deep in the Soviet Union.

Two versions of the carrier aircraft existed. The first had a conventional fuselage. The second used a twin boom arrangement, as in the Gotha Go 242 cargo glider or the American Lockheed P-38 *Lightning* fighter.

Each of these arrangements would carry different loads. The variant known as *Projekt A*, consisting of a carrier aircraft with a simple fuselage and a twin-engine jet bomber will be described here.

Two types of propulsion were envisaged, either HeS 021 propeller turbines or HeS 011 turbojets of equivalent power. These would be mounted on pylons above the wing surfaces. This system of mounting the engines was, by itself, visionary because it anticipated the mounting of jet engines on pylons under the wings of later jet aircraft.

In contrast to *Mistel* combinations, the bomber (known as *Projekt F*) would be carried beneath the carrier, suspended between the tall legs of the fixed landing gear, whose track was 25 meters.

The bomber showed a very advanced configuration, with wings swept at 38°, two turbojets mounted below them, and a "V" tail to facilitate installation beneath the carrier plane. It would be carried to the neighborhood of the target and launched, covering the final distance at high speed before delivering its load of up to 30,000 kg of bombs. The carrier would turn back and return safely to its base.

Used in an attack on the United States, the bomber would not have the ability to return to Europe. It would instead go to a pre-determined point in the Atlantic Ocean and ditch. If the crew survived they would patiently wait for the arrival of a German submarine to rescue them.

The concept of a carrier aircraft was used later in the United

Luftwaffe - confidential - Fundamentals of modern aeronautical design

A majestic view of the Shuttle Carrier in flight. Photo: NASA.

States in project FICON (*Fighter Conveyor*). This was the combination of a B-36 Bomber and a modified version of the RF-84 reconnaissance airplane, known as RF-84K *Thunderflash*.

In the FICON system, the small recon airplane would be carried by the B-36 until near the target. After release, it would fly at high speed in order to penetrate heavily defended targets and carry out its reconnaissance run.

Once the mission was accomplished, the RF84K would return to the B-36 which would have remained outside the reach of enemy defenses. The recon airplane would then be brought back on board the B-36 using a trapeze similar to that employed in the unsuccessful *Goblin* project (see PARASITE FIGHTERS).

In contrast to the XF-85, Project FICON was successful and in 1955, reached operational status in the Strategic Air Command.

In more recent times, the *Mistel* concept found another application in the so called Shuttle Carrier Aircraft (SCA). This is a specially modified Boeing 747 plane, configured to carry the Space Shuttle above the 747's fuselage.

These combinations were initially used to perform the first gliding flights of the Shuttle and were also used to transport the spacecraft from landing sites back to their launching base.

The same solution was adopted by the Soviet Union in 1989. To transport their version of the Space Shuttle, known as *Buran*, it was mounted above the fuselage of the gigantic Antonov 225 *Mryia*.

Spy Aircraft

DFS 228

The DFS 228 used typical glider construction techniques with the exception of its pressurized cabin, which was entirely of metal. The wings, tail and aft sections of the fuselage, were all manufactured in wood. For the operational versions, the installation of a Zeiss camera was foreseen. The DFS 228 would land on a metal skid, installed under the central section of the fuselage.

Data File

Mission:	Experimental and reconnaissance aircraft
Final development status:	prototype
Engine:	one Walter rocket engine
Maximum speed:	900 km/h (powered flight)
Range:	1,050 km (combined powered and glide flight)
Weight:	3,775 kg (loaded)
Span:	17.40 meters
Length:	10.12 meters
Armament:	---------------

DFS 346

The DFS 346 was entirely manufactured in metal. The Soviet version incorporated modifications to the German design, including a longer and more slender fuselage. The wings had wing fences of the same type as those adopted by the MiG 15. When returning from its mission, the airplane would land on a retractable skid housed in the fuselage.

Data File

Mission:	Experimental aircraft
Final development status:	Prototype (manufactured in Russia)
Engine:	two Walter rocket engine
Maximum speed:	2,011 km/h (theoretical)
Range:	8 minutes endurance
Weight:	8,000 kg (loaded)
Span:	8.99 meters
Length:	11.65 meters
Armament:	---------------

Parasite Airplanes

Me P - 1073

MESSERSCHMITT P-1073A

Project P-1073A is one of the first applications of the swept wings concept (35°) by Messerschmitt. The drawing shows a set of skis, probably intended to allow landing on the water since the recovery system had not been completely designed, and the use of a towing cable was suggested as an alternative.

Data File

Mission:	Parasite fighter
Final development status:	design
Engine:	one BMW 3304 turbojet engine
Maximum speed:	935 km/h
Range:	Approx. 31 minutes flight at 50% thrust
Weight:	1,620 kg (loaded)
Span:	4.40 meters
Length:	5.90 meters
Armament:	Two Mg 151/20 cannons

Rocket Airplanes

Sänger

The *Silbervogel* was to be entirely built in metal. In order to solve the problem of heating caused by aerodynamic friction, a refrigeration system based on a sheet of water and evaporation was foreseen. The cockpit cabin would be pressurized and cooled by air-conditioning. When comparing these solutions with the complex systems of thermal insulation used in the Space Shuttle, it must be concluded that the *Silbervogel* was an impractical proposal with the state of technological development as it was in Germany in 1945.

The metallic fuselage presented a profile similar to that of a wing, with the bottom flat and the upper surface curved, thus generating part of the aerodynamic lift. This same arrangement was later tried by the X-23 and X-24 lifting-body aircraft. The wings had a wedge profile and the twin tail had directional rudders. Some drawings show the installation of a conventional wheeled landing gear, but the use of landing skids is also mentioned.

Data File

Mission:	Sub orbital bomber
Final development status:	Wind tunnel models
Engine:	one Sänger rocket engine
Maximum speed:	22,100 km/h (theoretical)
Range:	23,500 km
Weight:	200,000 kg (loaded)
Span:	15.00 meters
Length:	28.00 meters
Armament:	The bomb load varied between one to five tons

Miniature Fighters

BV P-213

Data File

Mission:	Miniature fighter
Final development status:	design
Engine:	one Argus As 014 pulse-jet engine with 300 kg thrust
Maximum speed:	650 km/h
Range:	170 km
Weight:	1,280 kg (loaded)
Span:	6.00 meters
Length:	6.20 meters
Armament:	One Mk 108 cannon

IIn November 1944, the RLM circulated the requirement for an extremely simple fighter that should have been built even more easily than the Heinkel He 162 *Volksjäger*. Three companies, Blohm und Voss, Heinkel, and Junkers presented proposals for the *Miniaturjäger*, as it became known. Blohm und Voss' proposal anticipated a minimal use of strategic raw materials and labor. The BV P 213 was a minuscule high wing monoplane, powered by an Argus As 014 pulse-jet of the kind used in the V1 flying bombs. In order to save aluminum, not only the wings but also the inverted "V" empennage were made of wood. The steel fuselage was built from two welded halves. Its armament would consist of a single MK 108 cannon.

Mistel - Combined Aircraft

Daimler Benz Project A+F

Data File

Mission:	Carrier aircraft
Final development status:	design
Engine:	Two PTL 021 turboprop engines
Maximum speed:	640 km/h
Range:	6,500 km
Weight:	53,000 kg (only the carrier)
Span:	54.00 meters
Length:	35.00 meters
Armament:	----------------

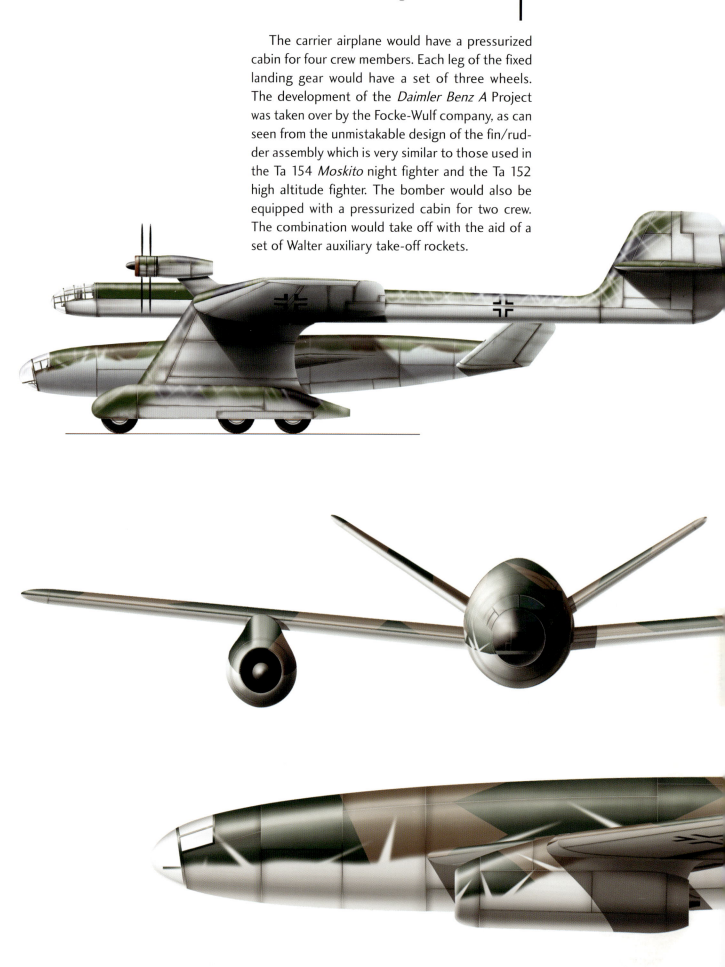

The carrier airplane would have a pressurized cabin for four crew members. Each leg of the fixed landing gear would have a set of three wheels. The development of the *Daimler Benz A* Project was taken over by the Focke-Wulf company, as can be seen from the unmistakable design of the fin/rudder assembly which is very similar to those used in the Ta 154 *Moskito* night fighter and the Ta 152 high altitude fighter. The bomber would also be equipped with a pressurized cabin for two crew. The combination would take off with the aid of a set of Walter auxiliary take-off rockets.

Data File

Mission:	bomber
Final development status:	design
Engine:	two BMW 018 turbojet engines each with 6,500 kg thrust
Maximum speed:	918 km/h
Range:	2,500 km
Weight:	71,800 kg (loaded)
Span:	26.16 meters
Length:	----------------
Armament:	four Mg 151/20 cannons fixed to fire backwards and up to 30,000 kg of bombs

SPECIAL MISSIONS

BIBLIOGRAPHY

1. ANDRADE, Roberto Pereira; PIOCHI, Antônio Ermete. **História da Construção Aeronáutica No Brasil.** Santana: Aquarius Editora, 1982.

2. BOYNE, Walter. **Boeing B-52 - A documentary history.** London: Jane's Publishing Company, 1981.

3. BURZACO, Ricardo. **Las alas de Peron II - La Aeronáutica Argentina 1945-1960.** Buenos Aires: Ediciones Eugenio Banfield, 2007.

4. COATES, Steve. **Helicopters of the Third Reich.** Surrey: Ian Allan Publishing, 2002.

5. COLBY, C. B. **Jets of the World.** New York: Coward-McCann, 1952.

6. CREEK, Eddie J.; SMITH, John Richard. **Jet Planes of the Third Reich.** Boylston: Monogram Aviation Publications, 1982.

7. CREEK, Eddie J.; SMITH, John Richard. **Arado 234 Blitz.** Sturbridge: Monogram Aviation Publications, 1992.

8. CREEK, Eddie J.; FORSYTH, Robert. **Heinkel He 162 Spatz.** Hersham: Ian Allan Publishing, 2008.

9. CROSS, Roy; GREEN, William. **The Jet Aircraft of the World.** London: Macdonald, 1956.

10. DABROWSKI, Hans-Peter. **Lippisch P13a & Experimental DM-1.** Atglen: Schiffer Publishing Ltd., 1991.

11. DABROWSKI, Hans-Peter. **Nurflügel.** Wölfersheim-Berstadt: Podzum-Pallas Verlag, 1996.

12. DIEDRICH, Hans-Peter. **German rocket fighters of World War II.** Atglen: Schiffer Military History, 2005.

13. DRESSEL, Joachim; GRIEHL, Manfred. **Die deutschen Raketenflugzeuge 1935-1945.** Stuttgart: Motorbuch Verlag, 1989.

14. DRESSEL, Joachim; GRIEHL, Manfred. **Deutsche Hubschrauber vor 1945.** Friedberg: Podzun-Pallas Verlag, 1991.

15. EBERT, Hans J.; KAISER, Johann B.; PETRES, Klaus. **Willy Messerschmitt: Pioneer of Aviation Design.** Atglen: Schiffer Publishing, Ltd., 1999.

16. FORSYTH, Robert. MISTEL: **German Composite Aircraft and Operations 1942-1945.** Crowborough: Classic Publications Limited, 2001.

17. GILL, Anton; HYLAND, Gary. **Last talons of the eagle.** London: Headline Book Publishing, 1998.

18. GINTER, Steve. **Convair XFY-1 Pogo.** Simi Valley: Ginter Books, 1994.

19. GINTER, Steve. **Lockheed XFV-1 VTOL Fighter**. Simi Valley: Ginter Books, 1996.

20. GORDON, Yefim. **Early Soviet Jet Bombers**. Hinckley: Midland Publishing, 2004.

21. GORDON, Yefim. **Soviet Rocket Fighters**. Hinckley: Midland Publishing, 2006.

22. GORDON, Yefim; KOMISSAROV, Dmitry. **German aircraft in the Soviet Union and Russia**. Hinckley: Midland Publishing, 2008.

23. GRALLERT, Herbert; KOELLE, Dietrich E.; SACHER, Peter. **Deutsche Raketenflugzeug und Raumtransporter–Projekte**. Bonn: Bernard & Graefe, 2007.

24. GREEN, William; POLLINGER, Gerald. **The Aircraft of the World.** London: Macdonald, 1956.

25. GREEN, William. **Warplanes of the Third Reich**. London: Macdonald and Jane's, 1979.

26. GRIEHL, Manfred. **Jet planes of the Third Reich, the secret projects vol. 1**. Sturbridge: Monogram Aviation Publications, 1998.

27. GRIEHL, Manfred. **Jet planes of the Third Reich, the secret projects vol.2**. Sturbridge: Monogram Aviation Publications, 2004.

28. HERWIG, Dieter; RODE, Heinz. **Geheimprojekte der Luftwaffe: Strategische Bomber 1935-1945**. Stuttgart: Motorbuch Verlag, 1998.

29. HERWIG, Dieter; RODE, Heinz. **Luftwaffe secret projects: Ground attack & special purpose aircraft**. Hinckley: Midland Publishing Company, 2003.

30. HORTEN, Reimar; SELINGER, Peter F. **Nürflugel: Die Geschichte der Horten Flugzeuge 1933-1960**. Graz: Herbert Weishaupt Verlag, 1983.

31. KHAZANOV, D. B., SOBOLEV, D. A. **The german imprint on the history of russian aviation**. Moscow: RUSAVIA, 2001.

32. LIPPISCH, Alexander. **Ein Dreieck fliegt ; Die Entwicklung der Delta-Flugzeuge**. Suttgart: Motorbuch Verlag, 1976.

33. LLOYD, Alwyn T. **B-47 Stratojet**. Blue Ridge Summit: TAB Books, 1986.

34. LOMMEL, Horst. **Geheimprojekte der DFS - Vom Höhenaufklärer bis zum Raumgleiter 1935-1945**. Stuttgart: Motorbuch Verlag, 2000.

35. LOMMEL, Horst. **The Junkers Ju 287 - The world's first swept-wing jet aircraft**. Atglen: Schiffer Publishing, Ltd., 2004.

36. MASTERS, David. **German Jet Genesis**. Londron: Jane's Publishing Company, 1982.

37. MATHEWS, Henry. **Samolyot 346**. Beirut: HPM Publications, 1996.

38. MEYER, Ingolf; SCHICK, Walter. **Luftwaffe secret projects: fighters 1939-1945**. Earl Shilton: Midland Counties Publications, 1997.

39. MEYER, Hans-Ulrich. **Die Pfeilflügelentwicklung in Deutschland bis 1945**. Bonn: Bernard & Graefe Verlag, 2006.

40. MILLER, Jay. **Convair B-58**. Arlington: AEROFAX Inc., 1985.

41. MILLER, Jay. **The X-planes X-1 to X-45**. Hinckley: Midland Publishing, 2001.

42. NOWARRA, Heinz J. **Die Deutsche Luftrüstung 1933-1945 Band 1-4**. Koblenz: Bernard & Graefe Verlag, 1988.

43. NOWARRA, Heinz J. **Focke-Wulf Fw 190 & Ta 152 - Aircraft and Legend**. Sparkford: Haynes Publishing Group, 1988.

44. OTTENS, Huib; SHEPELEV, Andrei. **Horten Ho 229 - Spirit of Thuringia**. Hersham: Ian Allan Publishing, 2006.

45. PELLETIER, Alain. **Les ailes Volants**. Boulogne: E-T-A-I, 1999.

46. POHLMANN, Hermann. **Chronik eines flugzeugwerkes**. Stuttgart: Motorbuch Verlag, 1982.

47. RADINGER, Willy; SCHICK, Walter. **Secret Messerschmitt projects**. Atglen: Schiffer Publishing Ltd., 1996.

48. RANSOM, Stephen. **Junkers Ju 287 - Germany's Forward Swept Wing Bomber**. Hersham: Ian Allan Publishing, 2008.

49. ROSE, Bill. **Secret projects - Flying wings and tailless aircraft**. Hersham: Ian Allan Publishing, 2010.

50. SPRINGMANN, Enno. **Focke - Flugzeuge und Hubschrauber von Henrich Focke 1912-1961**. Oberhaching: Aviatic Verlag, 1997.

51. STORCK, Rudolf. **Flying wings**. Bonn: Bernard & Graefe Verlag, 2003.

52. STURTIVANT, Ray. **British Prototype Aircraft**. London: Promotional Reprint Company, 1995.

53. WAGNER, Wolfgang. **The first jet aircraft**. Atglen: Schiffer Publishing Ltd., 1998.

54. WOOLDRIDGE, E. T. **Winged Wonders**. Washington: National Air and Space Museum, 1985.

Photo by Carsten Selinger

Authors

Claudio Lamas de Farias (short guy)

Was born in 1960, in Rio de Janeiro. Graduated as an industrial designer in 1982. Obtained a Master Degree in Design with a dissertation about the history of Airliner Interior Design. Today teaches design history at the Federal University of Juiz de Fora, in the state of Minas Gerais.

Daniel Uhr (tall guy)

Was born in 1968, in Rio de Janeiro.
Graduated as a graphic designer in 1994.
Obtained a MBA in Strategic Marketing in 1999.
Has done many illustrations for magazines such as Força Aérea (Brazil), Aviation History (USA), Ian Allan Publishing (UK) and Fluid (Poland) and also for the Luft46 website. Also designed box covers for plastic kits companies such as Airmodel (Germany), Hobby Boss, Trumpeter and AniGrand (China), Valom (Czech Republic) and HighPlane (Australia), among others. Opened his own design studio, Eletronicmedia (www.eletronicmedia.com.br) in 2002. Later created a division devoted entirely to aviation art - D.Uhr Aviation Art (www.duhraviationart.com). His aviation art was displayed in an individual exhibition called "Impressions of a dream", initially in Rio de Janeiro in 2008 and then travelled to Blumenau, where it stayed for three months, and Petropolis. Today teaches (graphic design, web design and scale model) at SENAC, in Rio de Janeiro.